Handbook of Radio Frequency Identification

Handbook of Radio Frequency Identification

Edited by **Kevin Merriman**

CLANRYE
INTERNATIONAL

New Jersey

Published by Clanrye International,
55 Van Reypen Street,
Jersey City, NJ 07306, USA
www.clanryeinternational.com

Handbook of Radio Frequency Identification
Edited by Kevin Merriman

International Standard Book Number: 978-1-63240-286-8 (Hardback)

Printed in the United States of America.

Contents

Preface

This book has been a concerted effort by a group of academicians, researchers and scientists, who have contributed their research works for the realization of the book. This book has materialized in the wake of emerging advancements and innovations in this field. Therefore, the need of the hour was to compile all the required researches and disseminate the knowledge to a broad spectrum of people comprising of students, researchers and specialists of the field.

Radio Frequency Identification (RFID) is an emerging data sharing technology. This book elucidates issues related to various aspects of RFID technology, its applications and architectural patterns. RFID based functions create incredible new business opportunities such as the support of autonomous living of aged and disabled persons, resourceful supply chains, and well-organized anti-counterfeiting and improved ecological monitoring. Its data administration, scalable information systems, business procedure reengineering and evaluating reserves are emerging as important technological challenges to applications underpinned by new advancements in RFID technology. It presents researches from prominent professionals on the newest developments in the RFID field.

At the end of the preface, I would like to thank the authors for their brilliant chapters and the publisher for guiding us all-through the making of the book till its final stage. Also, I would like to thank my family for providing the support and encouragement throughout my academic career and research projects.

Editor

An RFID Anti-Collision Algorithm Assisted by Multi-Packet Reception and Retransmission Diversity

Ramiro Sámano-Robles, Neeli Prasad and
Atílio Gameiro

Additional information is available at the end of the chapter

1. Introduction

1.1. RFID technology and previous works

RFID (Radio Frequency Identification) is a technology that uses radio frequency signals for purposes of identification and tracking of objects, humans or animals. In passive RFID systems, where tags reuse the energy radiated by the reader, coordination capabilities can be considerably limited [29]. This issue leads to conflicts or collisions between the transmissions of the different elements of an RFID network, i.e., readers an tags. An efficient medium access control layer (MAC) is thus crucial to the correct operation of RFID [3].

Two types of RFID MAC collision can be identified: tag and reader collision. A tag collision arises when several tags simultaneously respond to a given reader request, thus causing the loss of all the transmitted information. To address this issue, tag anti-collision schemes such as ALOHA and binary tree algorithms are commonly employed [3, 31]. Improvements on these solutions have been further proposed by using tag estimation methodologies [14], and modified frame structures [3, 30], among many other approaches in the literature. Two types of reader collision can also be identified: multiple-reader-to-tag collision and reader-to-reader collision [2]. To address these two issues, reader anti-collision algorithms based on scheduling or coverage control have been proposed. Typical scheduling schemes are frequency division multiple access (FDMA) [7] or listen-before-talk (LBT) [8]. Advanced schemes such as Colorwave in [28] and Pulse in [2] implement inter-reader control mechanisms to assist in the collision resolution process. Other approaches such as HiQ in [11] use an analysis of collision patterns over consecutive time-slots to improve scheduling policies. Regarding coverage-based algorithms, two types of scheme can be commonly found: those that reduce the overlapping coverage area between readers (e.g., [12]), and those that monitor interference to adapt power levels accordingly (e.g., [4]).

1.2. Open issues and chapter objectives

Despite recent advances in RFID MAC layer design, several issues remain open today. Current RFID algorithms are designed under simplistic assumptions such as the collision-model. In such a collision-model, collisions are regarded as the loss of all the transmitted information. On the contrary, collision-free transmissions are always assumed to be correctly received. These assumptions are, however, highly inaccurate, particularly for wireless settings with rapidly changing channel conditions and assisted by modern signal processing tools. In wireless networks, packet transmissions can be lost due to random fading phenomena and not only due to collisions. On the other hand, a collision with multiple concurrent transmissions can be resolved by means of multiple antenna receivers. Therefore, a new approach for a more accurate design and modeling of random access protocols in modern wireless networks is required. In the literature of conventional random access protocols, considerable advances in these aspects have been recently made using the concept of cross-layer design [18–26]. The objective of this chapter is to use two of these recent cross-layer solutions and modeling approaches to improve the performance of RFID. In particular, we focus on those solutions that make use of signal processing tools that exploit diversity in the space (multi-packet reception) and time domains (retransmission diversity).

1.3. MAC-PHY cross-layer design: Multi-packet reception and retransmission diversity

Multi-packet reception is a concept that has revolutionized the design paradigm of random access protocols. Conventionally, collisions were always considered as the loss of all the transmitted information. However, modern multiuser detection and source separation tools allow for the simultaneous decoding of concurrent transmissions. Design of random access protocols with multi-packet reception has been addressed in [9] using a symmetrical and infinite user population model, and in [16] using an asymmetrical and finite user population model. A novel multi-packet reception scheme that exploits the time domain in order to achieve diversity has been proposed in [26], and it has been called network diversity multiple access (NDMA). In NDMA, a virtual MIMO (multiple-input multiple-output) system is induced by requesting as many retransmissions as needed to recover the contending packets using source separation. A hybrid algorithm with multi-packet reception and retransmission diversity has been proposed in [21].

1.4. Chapter contributions

This chapter aims to use the concepts of multi-packet reception and retransmission diversity in the MAC layer design of passive RFID systems. To investigate these two cross-layer random access algorithms in the context of RFID, a novel framework which includes PHY (physical) and MAC (medium access control) layer parameters of RFID is here employed. The framework consists of the co-modeling of both the down-link (reader-to-tag) and up-link (tag-to-reader) signal-to-interference-plus-noise ratio (SINR) experienced in a multi-tag and multi-reader environment. This framework was first proposed in our previous work in [22], and it has been modified here to be used in the context of multi-packet reception and retransmission diversity. Based on this updated framework, stochastic models for tag activation/detection processes (considering multi-packet reception and retransmission diversity) are then proposed. The proposed approach also allows for a novel joint design of reader and tag anti-collision schemes. Conventionally, these two algorithms were designed independently from each other. However, readers and tags operate in the same frequency band. Therefore, contention between their transmissions can potentially arise. Furthermore, reader anti-collision policies directly influence tag activation, and thus also the way in which tags collide when responding to readers' requests. Therefore, a complete model of RFID MAC layer should consider both processes together rather than independently. The proposed framework fills this gap by simultaneously modeling tag activation and the corresponding tag responses to readers, while also considering multi-packet reception and retransmission diversity at the reader side.

To complement the framework for MAC/PHY cross-layer design, a Markov model is also presented, which allows for capacity and stability evaluation of asymmetrical RFID systems. The approach consists of defining the states (i.e., the set of active tags/readers) that describe the network at any given time, and then map them into a one-dimensional Markov model that can be solved by standard techniques such as eigenvalue analysis. The results show that the proposed algorithms as well as the joint cross-layer approach and the Markov model provide considerable benefits in terms of capacity and stability over conventional solutions.

1.5. Organization

The organization of this chapter is as follows. Section 2 describes the framework for cross-layer design, and gives details of the operation of the protocol with multi-packet reception and retransmission diversity. Section 3 describes the proposed metrics and the Markov model. Section 4 addresses the optimization of the system and displays the results using different scenarios. Finally, Section 5 presents the conclusions of the chapter.

2. System model and cross-layer framework

Consider the slotted RFID network depicted in Fig. 1 with a set $\mathcal{R}$ of K readers $\mathcal{R} = \{1, \dots K\}$ and a set $\mathcal{T}$ of J tags $\mathcal{T} = \{1, \dots, J\}$. Each reader is provided with M antennas that will be used to recover, using source separation, the simultaneous transmissions of several tags. Two main processes can be distinguished in the RFID network in Fig. 1: Tag activation by the transmission of readers, also called the down-link transmission; and the backscattering response towards readers by previously activated tags, also called up-link transmission. In the down-link, the transmit power of reader k will be denoted by $P_{r,k}$ while its probability of transmission will be denoted by $p_{r,k}$. All the antennas will be assumed to transmit the same signal in the down-link. The subset of active readers at any given time will be denoted by $\mathcal{R}_t$. Tags are activated when the signal-to-interference-plus-noise ratio (SINR) given a reader transmission is above an activation threshold. The set of activated tags will be denoted by $\mathcal{T}_P$. In the up-link, the active tags proceed to transmit a backscatter signal using a randomized transmission scheme. The subset of tags that transmit a signal once they have been activated will be given by $\mathcal{T}_t$, where each tag j will transmit with a power level denoted by $P_{t,j}$. Details of the down- and up-link models are given in the following subsections.

2.1. Tag activation process: Down-link model

Consider that the instantaneous channel between reader k and tag j is given by the column vector $\mathbf{h}_{k,j}$ with dimensions $M \times 1$, the channel experienced between reader k and reader m is given by the matrix $\mathbf{G}_{k,m}$ with dimensions $M \times M$, and the channel experienced between tag i and tag j is given by the scalar value $u_{i,j}$. The SINR experienced by tag j due to a transmission of reader k is denoted by $\gamma_{k,j}$, and it can be mathematically expressed as follows:

$$\gamma_{k,j} = \frac{P_{r,k}\mathbf{h}_{k,j}^H\mathbf{h}_{k,j}}{I_{r_{k,j}} + I_{t_j} + \sigma_{v,j}^2}, \qquad k \in \mathcal{R}_t, \tag{1}$$

where $I_{r_{k,j}} = \sum_{m \in \mathcal{R}_t, m \neq k} P_{r,m}\mathbf{h}_{m,j}^H\mathbf{h}_{m,j}$ is the interference created by other active readers, $I_{t_j} = \sum_{i \in \mathcal{T}_i, i \neq j} P_{t,i}(|u_{j,i}|^2)$ is the interference created by other tags, $(\cdot)^H$ is the hermitian transpose operator, and $\sigma_{v,j}^2$ is the noise. If the SINR experienced by tag j is above the tag sensitivity threshold $\tilde{\gamma}_j$, then the tag becomes activated. The probability of tag j, which was previously inactivated, to become activated will be given by

$$\Pr\{j \in \mathcal{T}_P\} = \Pr\{\max_k \gamma_{k,j} > \tilde{\gamma}_j\}. \tag{2}$$

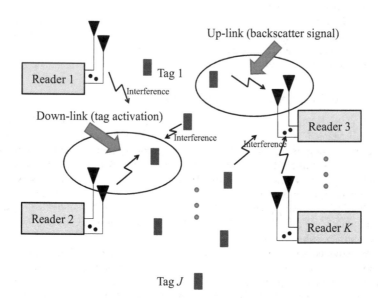

Figure 1. Multi-tag and Multi-reader deployment scenario.

2.2. Backscattering process: Up-link model

Once a tag j has been activated by the transmission of a given reader, it then starts a random transmission process to prevent collisions with other active tags using a Bernoulli process with parameter $p_{t,j}$, which is also the transmission probability. The backscattering factor β_j is the fraction of the received power reused by the tag to reply to the reader. Therefore, the transmit power of tag j can be calculated as $P_{t,j} = \beta_j P_{r,k} |h_{k_{opt},j}|^2$, where $k_{opt} = \arg\max_k \gamma_{k,j}$ denotes the reader that has previously activated the tag. At the reader side, source separation tools for multi-packet reception and retransmission diversity will be used. The proposed protocol consists of ensuring that the number of diversity sources is equal or larger than the number of contending tags so that the source separation technique is successful. For example, if 4 tags collide at a particular time-slot (see Fig. 2) and the reader is provided with only 2 antennas, then the system will request a retransmission from the contending tags in the following time slot. The reader will store all the signals collected during these 2 time-slots and will create a virtual MIMO system from which the signals of the contending tags can be estimated using multiuser detection. The array of stacked signals received at reader k across all r sources of diversity[1] is given by:

$$\mathbf{Y}_{r,k} = \mathbf{H}\mathbf{S} + \mathbf{I}_{r,k} + \mathbf{E}_{r,k} + \mathbf{V}_{r,k} \tag{3}$$

where $\mathbf{H}$ is the stacked version of all the channels of the contending tags, $\mathbf{S}$ is the stacked version of all the signals of the contending tags, $\mathbf{I}_{r,k}$ is the collected interference created by other active readers, $\mathbf{E}_{r,k}$ is the collected leaked signal power from the transmission chain, and $\mathbf{V}_{r,k}$ is the noise term. At the reader side, a multiuser receiver such as zero forcing (ZF) or minimum mean square error (MMSE) can be implemented. For example, the zero forcing receiver can be described as follows:

$$\widehat{\mathbf{S}} = \widehat{\mathbf{H}}^{-1}\mathbf{Y}_{r,k}, \tag{4}$$

where $\widehat{\mathbf{S}}$ is the array of estimated signals of the contending tags, and $\widehat{\mathbf{H}}$ is the estimated channel of the contending tags. Since the resolution of a collision may take place over a random number of time slots due to the retransmission diversity scheme, then we will denote this collision resolution period as an *epoch-slot* with a length denoted by the random variable l_{ep}.

For simplicity, it will be assumed that the performance of the multiuser receiver is described by the ability to correctly detect the presence of all the contending tags. This assumption has been used in the analysis of conventional NDMA protocols in [26]. In this assumption any detection error yields the loss of all the contending packets. Thus, it is possible to propose the detection SINR of tag j at reader k, denoted by $\widehat{\gamma}_{j,k}$, as follows:

$$\widehat{\gamma}_{j,k} = \frac{P_{t,j}\mathbf{h}_{k,j}^H\mathbf{h}_{k,j}}{\widehat{I}_{r,k} + P_{r,k}\eta_k + \widehat{\sigma}_{v,k}^2}, \qquad j \in \mathcal{T}_t \tag{5}$$

---

[1] the number of diversity sources is the total number of combinations of antenna elements and retransmissions

where $\widehat{I}_{r,k} = \sum_{m \neq k} \text{tr}(\mathbf{G}_{k,m}^H \mathbf{G}_{k,m})$ is the interference created by other active readers, $\text{tr}(\cdot)$ is the trace operator, η_k is the power ratio leaked from the down-link chain, and $\widehat{\sigma}_{v,k}^2$ is the noise. Note that tag-to-tag interference is not considered as an independent orthogonal training signal for each tag is used in each transmission for purposes of tag detection and channel estimation, which is also used in the original NDMA protocol in [26]. Thus, tag j can be detected by reader k if the received SINR is above a threshold denoted by $\widecheck{\gamma}_k$. The set of detected tags by reader k will be denoted by $\mathcal{T}_{D,k}$, thus the probability of tag j being in $\mathcal{T}_{D,k}$ will be given by

$$\Pr\{j \in \mathcal{T}_{D,k}\} = \Pr\{\widehat{\gamma}_{j,k} > \widecheck{\gamma}_k\}. \tag{6}$$

The set of correctly detected tags across all the readers will be simply given by $\mathcal{T}_D$, where $\mathcal{T}_D = \cup_k \mathcal{T}_{D,k}$. Since this detection process is prone to errors, we will use in this paper the same assumption used in the original paper for NDMA in [26] where tags are only correctly received at the reader side if all the contending tags are correctly detected and none of the remaining silent tags is incorrectly detected as active (i.e., false alarm). This means that correct tag reception for tag j only occurs when:

$$\Pr\{j \in \mathcal{T}_R\} = \Pr\{\mathcal{T}_D = \mathcal{T}_t\}, \quad \text{where} \quad j \in \mathcal{T}_t, \tag{7}$$

where $\mathcal{T}_R$ is the set of tags correctly received at the reader side. A tag that has transmitted to the reader side can be correctly detected with probability P_D, which can be defined as:

$$P_D = \Pr\{j \in \mathcal{T}_D | j \in \mathcal{T}_t\} = \sum_k \Pr\{\widehat{\gamma}_{j,k} > \widecheck{\gamma}_k | j \in \mathcal{T}_t\}, \tag{8}$$

and which can be read as the probability that tag j is correctly detected as active given it has transmitted a signal. Similarly, the probability of false alarm is given by:

$$P_F = \Pr\{j \in \mathcal{T}_D | j \notin \mathcal{T}_t\} = \sum_k \Pr\{\widehat{\gamma}_{j,k} > \widecheck{\gamma}_k | j \notin \mathcal{T}_t\}, \tag{9}$$

which can be read as the probability that tag j is incorrectly detected as active when it has transmitted no signal at all.

3. Performance metrics and Markov model

The main performance metric to be used in this chapter is the average tag throughput, which can be defined as the long term ratio of correct tag readings to the total number of time-slots used in the measurement. Before providing an expression for this metric, it is first necessary to define the network state information, as well as the tag activation and tag reception probability models, and the definition of the Markov model for the dynamic analysis of an RFID network.

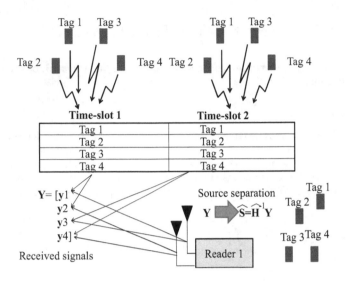

Figure 2. Example of the operation of the proposed protocol with multi-packet reception and retransmission diversity.

3.1. Network state information and tag activation model

The network state information can be defined as all the parameters that completely describe the network at any given time. In our case, the network state information $\mathcal{N}(n)$ at epoch-slot n is defined as the collection of the sets of active readers $\mathcal{R}_t(n)$ and contending tags $\mathcal{T}_t(n)$:

$$\mathcal{N}(n) = \{\mathcal{R}_t(n), \mathcal{T}_t(n)\}. \tag{10}$$

Once the network state information has been defined, we can define the probability of tag j being activated in slot n conditional on a given realization of the network state information $\mathcal{N}(n)$ and given that the tag was previously inactivated as follows:

$$Q_{j|\mathcal{N}(n)} = \Pr\{j \in \mathcal{T}_P(n+1)|\mathcal{N}(n), j \notin \mathcal{T}_P(n)\} = \Pr\{\max_k \gamma_{k,j}(n) > \tilde{\gamma}_j\}. \tag{11}$$

For convenience in the analysis, let us rewrite this tag activation probability in terms of the set of active tags $\mathcal{T}_P(n)$ by averaging over all values of $\mathcal{N}(n)$ where $\mathcal{T}_t(n) \in \mathcal{T}_P(n)$:

$$Q_{j|\mathcal{T}_P(n)} = \sum_{\mathcal{N}(n); \mathcal{T}_t(n) \in \mathcal{T}_P(n)} \Pr\{\mathcal{N}(n)\} Q_{j|\mathcal{N}(n)} \tag{12}$$

where $\Pr\{\mathcal{N}(n)\}$ is the probability of occurrence of the network state information $\mathcal{N}(n)$. This term can be calculated by considering all the combinations of active tags and readers as follows:

$$\Pr\{\mathcal{N}(n)\} = \prod_{k\in\mathcal{R}_t} p_{r,k} \prod_{m\notin\mathcal{R}_t} \overline{p}_{r,m} \prod_{j\in\mathcal{T}_t} p_{t,j} \prod_{i\notin\mathcal{T}_t} \overline{p}_{t,i} \qquad (13)$$

where $\overline{(\cdot)} = 1 - (\cdot)$. This concludes the definition of the tag activation probability and the network state information.

3.2. Markov model

In order to define the Markov model for dynamic analysis of the system, let us now calculate the probability of having a set of active tags $\mathcal{T}_P(n+1)$ in epoch-slot $n+1$ conditional on having the set of active tags $\mathcal{T}_P(n)$ during the previous epoch-slot. This transition probability must consider all the combinations of tags that either enter (i.e., they are activated in epoch slot n) or leave the set of active tags (i.e., they transmit in epoch slot n). This can be expressed as follows:

$$\Pr\{\mathcal{T}_P(n+1)|\mathcal{T}_P(n)\} = \prod_{j\in\mathcal{T}_P(n),j\notin\mathcal{T}_P(n+1)} p_{t,j} \prod_{i\notin\mathcal{T}_P(n),i\in\mathcal{T}_P(n+1)} Q_{i|\mathcal{T}_P(n)} \prod_{l\notin\mathcal{T}_P(n),l\notin\mathcal{T}(n+1)} \overline{Q}_{l|\mathcal{T}_P(n)}$$

$$\times \prod_{w\in\mathcal{T}_P(n),w\in\mathcal{T}_P(n+1)} \overline{p}_{t,w}. \qquad (14)$$

Let us now arrange the probability of occurrence of all the possible sets of activated tags $\Pr\{\mathcal{T}_P\}$ into a one-dimensional vector given by $\mathbf{s} = [s_0,\dots s_{JI}]^T$, where $(\cdot)^T$ is the transpose operator (see Fig. 3). This means that we are mapping the asymmetrical states into a linear state vector where each element represents the probability of occurrence of one different state $\Pr\{\mathcal{T}_P\}$. In the example given in Fig. 3 we have only two possible tags, where the first system state is given by both tags being active, the second state with only tag 1 as active, the third state with only tag 2 as active, and the fourth state with both tags inactive. Once these states are mapped into the state vector $\mathbf{s}$, the transition probabilities between such states ($\Pr\{\mathcal{T}_P(n+1)|\mathcal{T}_P(n)\}$) can also be mapped into a matrix $\mathbf{M}$, which defines the Markov model for state transition probabilities (see Fig. 3). The i,j entry of the matrix $\mathbf{M}$ denotes the transition probability between state i and state j. The vector of state probabilities can thus be obtained by solving the following characteristic equation:

$$\mathbf{s} = \mathbf{Ms}, \qquad (15)$$

by using standard eigenvalue analysis or iterative schemes. Each one of the calculated terms of the vector $\mathbf{s}$ can be mapped back to the original probability space $\Pr\{\mathcal{T}_P\}$, which can then be used to calculate relevant performance metrics.

3.3. Tag detection model

Before calculating the tag throughput, first we must define the correct reception probability of tag j at the reader side conditional on the network state information $\mathcal{N}(n)$ as follows:

$$q_{j|\mathcal{N}(n)} = \Pr\{j \in \mathcal{T}_R(n+1)\} = \Pr\{\mathcal{T}_D = \mathcal{T}_t\}, \quad \text{where} \quad j \in \mathcal{T}_t \tag{16}$$

It is also convenient to re-write this reception probability in terms of the set of active tags $\mathcal{T}_P(n)$ by averaging over all values of $\mathcal{N}(n)$ where $\mathcal{T}_t(n) \in \mathcal{T}_P(n)$:

$$q_{j|\mathcal{T}_P(n)} = \sum_{\mathcal{N}(n); \mathcal{T}_t(n) \in \mathcal{T}_P(n)} \Pr\{\mathcal{N}(n)\} q_{j|\mathcal{N}(n)} \tag{17}$$

3.4. Tag throughput and stability

The tag throughput per resolution period can be finally calculated by adding all the contributions over the calculated probability space $\Pr\{\mathcal{T}_P\}$ using the Markov model presented in previous subsections. This can be mathematically expressed as:

$$S_j = \sum_{\mathcal{T}_P, j \in \mathcal{T}_P} \Pr\{\mathcal{T}_P\} q_{j|\mathcal{T}_P}. \tag{18}$$

Now, the throughput per time-slot can be calculated as the ratio of the throughput per resolution period to the average number of time-slots per resolution period:

$$T_j = \frac{S_j}{\sum_{\mathcal{T}_D} \Pr\{\mathcal{T}_D\} \left\lceil \frac{|\mathcal{T}_D|}{M} \right\rceil + \Pr\{\mathcal{T}_D = \varnothing\}}, \tag{19}$$

where $|\cdot|$ is the set cardinality operator and $\lceil \cdot \rceil$ is the ceil integer operator. As a measure of stability we will use the average number of activated tags, which can be calculated as follows:

$$E[|\mathcal{T}_P|] = \sum_{\mathcal{T}_P} \Pr\{\mathcal{T}_P\} |\mathcal{T}_P|. \tag{20}$$

A high number of activated tags means that stability is compromised, while a relatively low number indicates that the algorithm is more stable.

4. Optimization and results

The parameters to be optimized are the vector of reader transmission probabilities $\mathbf{p}_r = [p_{r,1} \ldots p_{r,K}]^T$, the vector of reader transmit powers $\mathbf{P}_r = [P_{r,1} \ldots P_{r,K}]$ and the vector of transmission probabilities of the active tags $\mathbf{p}_t = [p_{t,1} \ldots p_{t,J}]$. The objective of the optimization is the total throughput, so the optimization problem with transmit power constraint can thus be written as follows:

$$\{\mathbf{P}_r, \mathbf{p}_t, \mathbf{p}_r\}_{opt} = \arg \max_{\{\mathbf{P}_r, \mathbf{p}_t, \mathbf{p}_r\}} \sum T_j \quad \text{s.t.} \quad \mathbf{P}_r < \mathbf{P}_{r,0} \tag{21}$$

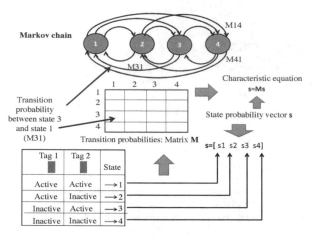

Figure 3. Example of the Markov model for a two-tag system.

Since the explicit optimization of the expressions is difficult to achieve, particularly when considering the Markov model proposed in the previous section, in this section we will simplify the optimization problem by applying the previous concepts to an ALOHA protocol implemented at the reader side. This means that tags can be only activated when the readers' transmissions are collision-free. Power levels will be fixed, and the maximum throughput performance will be investigated by simply plotting the surface versus the reader and tag transmission probabilities. At the tag side we will consider the following three options: a conventional ALOHA protocol without MPR, ALOHA with multi-packet reception (simply tagged ALOHA MPR), and the proposed scheme with retransmission diversity and multi-packet reception (tagged NDMA MPR).

Two scenarios are considered: one in which tags and readers operate in the same channel, thereby interfering with each other, and the second scenario where readers and tags operate in a synchronized manner in different channels, which eliminates the probability of collision between them. For convenience, let us consider in first instance that all tags and readers experience channel and queuing states that are statistically identical (symmetrical system). A tag activation probability of $q = 0.7$ and a tag detection probability at the reader side of $Q = 0.95$ have been used in the theoretical calculations. A probability of false alarm for the NDMA protocol has been set to $P_F = 0.01$. The results have been calculated with $J = 15$ tags and $K = 5$ readers.

Fig. 4 shows the results of average throughput $T = \sum_j T_j$ versus various values of reader and tag transmission probability p_t and p_r for a conventional ALOHA protocol without multi-packet reception ($M = 1$) and without retransmission diversity considering full interference between readers an tags. Fig. 5 shows the results of average throughput T versus various values of reader and tag transmission probability p_t and p_r for a conventional ALOHA protocol without multi-packet reception ($M = 1$) and without retransmission diversity considering no interference between readers an tags. Note how the throughput shape is considerably affected by the interference assumption between readers and tags. Fig. 6 shows the results of average number of tags versus various values of reader and tag transmission probability p_t and p_r. Fig. 7 shows the results of average throughput T

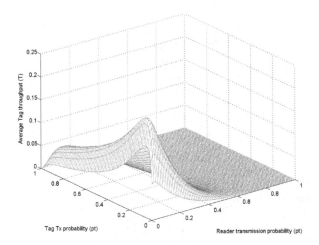

Figure 4. Throughput (T) vs. reader and tag transmissions probabilities (p_r and p_t) of a symmetrical ALOHA protocol for reader and tag anti-collision assuming interference between readers and tags.

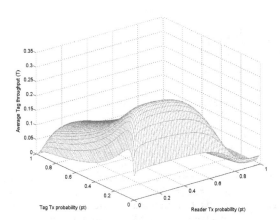

Figure 5. Throughput (T) vs. reader and tag transmissions probabilities (p_r and p_t) of a symmetrical ALOHA protocol for reader and tag anti-collision assuming no interference between readers and tags.

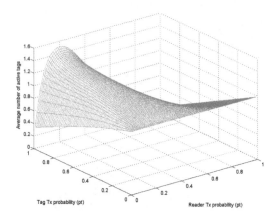

Figure 6. Average number of active tags vs. reader and tag transmissions probabilities (p_r and p_t).

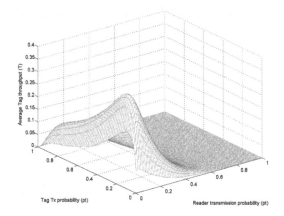

Figure 7. Throughput (T) vs. reader and tag transmissions probabilities (p_r and p_t) of a symmetrical ALOHA MPR protocol for reader and tag anti-collision assuming interference between readers and tags.

versus various values of reader and tag transmission probability p_t and p_r for a conventional ALOHA protocol with multi-packet reception ($M = 2$) and without retransmission diversity considering no interference between readers an tags. Note that the maximum throughput has been considerably improved over the conventional ALOHA protocol without MPR capabilities in Fig. 4. Similarly, Fig. 8 shows the results of average throughput T versus various values of reader and tag transmission probability p_t and p_r for a conventional ALOHA protocol with multi-packet reception ($M = 2$) and without retransmission diversity by considering no interference between readers an tags. The improvement over the algorithm without MPR in fig. 5 is considerable from almost 0.2 tags/timeslot in the case of ALOHA, to almost 0.4 tags/time-slot in the case of ALOHA MPR, and up to 0.7 tags/time-slot in the case of NDMA MPR. Fig. 9 shows the results of average throughput T versus various values

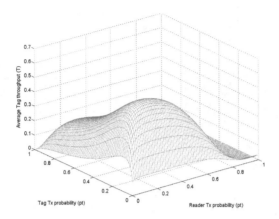

Figure 8. Throughput (T) vs. reader and tag transmissions probabilities (p_r and p_t) of a symmetrical ALOHA MPR protocol for reader and tag anti-collision assuming no interference between readers and tags.

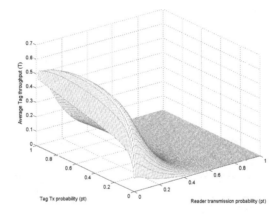

Figure 9. Throughput (T) vs. reader and tag transmissions probabilities (p_r and p_t) of a symmetrical NDMA MPR protocol for reader and tag anti-collision assuming interference between readers and tags.

of reader and tag transmission probability p_t and p_r for an NDMA MPR protocol considering interference between readers an tags, while Fig. 10 shows the results without considering interference between readers an tags. The results in Fig.9 and 10 show that the proposed NDMA MPR solution considerably outperforms its ALOHA counterparts in both scenarios: with or without interference between readers and tags.

Let us now address an asymmetrical scenario. For this purpose consider that the tag/reader space is divided into two different sets of readers and three different sets of tags. Readers and tags are working in different channels. The first and second sets of tags can only be reached by the first and second sets of readers, respectively. The third set of tags can be reached by both sets of readers. All tags have the same transmission probability p_t as well as

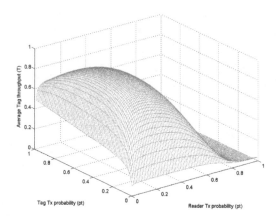

Figure 10. Throughput (T) vs. reader and tag transmissions probabilities (p_r and p_t) of a symmetrical NDMA MPR protocol for reader and tag anti-collision assuming no interference between readers and tags.

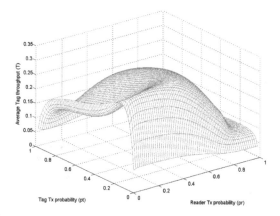

Figure 11. Throughput (T) vs. reader and tag transmissions probabilities (p_r and p_t) of an asymmetrical ALOHA protocol for reader and tag anti-collision without interference between readers and tags.

all readers transmit with the same parameter p_r. A tag activation probability of $q = 0.7$ and a tag detection probability at the reader side of $Q = 0.95$ have been used in the theoretical calculations. A probability of false alarm for the NDMA protocol has been set to $P_F = 0.01$. The results of Fig. 11 and Fig. 12 have been obtained using three groups of tags with $J_1 = 3, J_2 = 5$ and $J_3 = 7$ tags, and two groups of readers with $K_1 = 5$ and $K_2 = 10$ readers. While Fig. 11 shows the results of an ALOHA protocol without MPR capabilities ($M = 1$), Fig. 12 shows the results of the proposed NDMA protocol with $M = 1$. In both cases, the readers and tags are assumed to transmit in different channels, thereby avoiding interference between their transmissions. It can be observed the significant gain provided by the NDMA protocol for all values of p_t and p_r.

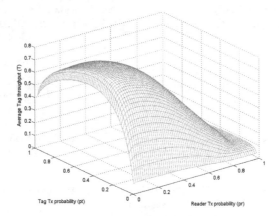

Figure 12. Throughput (T) vs. reader and tag transmissions probabilities (p_r and p_t) of an asymmetrical NDMA protocol for reader and tag anti-collision without interference between readers and tags.

5. Conclusions

This chapter presented a novel algorithm for passive RFID anti-collision based on the concepts of multi-packet reception and retransmission diversity. In addition, the design of the algorithm has been based on a new design paradigm called cross-layer design, where physical and medium access control layers are jointly designed, and where reader and tag anti-collision components are also jointly considered. The proposed Markov model is a new approach for the modeling of RFID networks, as it captures both the activation process given by the operation of readers sending requests to tags, and the tag detection process that results from tags randomly transmitting their information back to the readers that previously activated them. The results for tag throughput showed considerable improvement over conventional ALOHA solutions that have been implemented in current deployments and commercial platforms for RFID. This opens an interesting area for the design of advanced random access protocols for future RFID systems and for the internet of things.

Author details

Ramiro Sámano-Robles[1], Neeli Prasad[2] and Atílio Gameiro[1]

1 Instituto de Telecomunicações, Campus Universitário, Aveiro, Portugal
2 Center for TeleInfrastruktur, Department of Electronic Systems, Aalborg University, Aalborg, Denmark

References

[1] Ahmed, N.; Kumar, R.; French R.S. & Ramachandran, U.; (2007). RF2ID: A Reliable Middleware Framework for RFID Deployment. *IEEE International Parallel and Distributed processing Symposium*, March 2007, pp. 1-10.

[2] Birari S.M. & Iyer S. (2005). Mitigating the Reader Collision Problem in RFID Networks with Mobile Readers, *13th IEEE International Conference on Networks*, 2005, pp. 463-468

[3] Burdet L.A. (2004) RFID Multiple Access Methods, *IEEE Antennas and wireless propagation letters*, ETH Zürich, Summer semester 2004, Seminar Smart Environments.

[4] Cha K.; Jagannathan S. & Pommerenke D. (2007). Adaptive power control with hardware implementation for wireless sensor and RFID networks, *IEEE Systems Journal*, Vol. 1, No. 2, December 2007, pp. 145-159.

[5] Darianian, M.& Michael, M.P. (2008). Smart home Mobile RFID-based Internet-of-Things systems and services. *International Conference on Advanced Computer Theory and Engineering ICACTE*, 2008, pp.116-120.

[6] Eom J.B.;Yim S.B. & Lee T.J. (2009). An Efficient Reader Anticollision Algorithm in Dense RFID Networks With Mobile RFID Readers. *IEEE Transactions on Industrial Electronics*, Vol. 56, No. 7, July 2009, pp. 2326-2336.

[7] EPC (2009). "*EPC Radio-Frequency Identity Protocols. Class-1 Generation-2 UHF RFID. Protocol for communications at 860 MHz - 960Mhz Version1.2.0*", EPC Global, 2008., http://www.epcglobalinc.org/standards/.

[8] ETSI (2004). "*ETSI EN 302 208-1,2 v1.1.1*", September 2004. CTAN: http://www.etsi.org

[9] Ghez S.; Verdu S. & Schwartz S. (1988).Stability properties of slotted Aloha with multipacket reception capability, *IEEE Trans. on Aut. Cont.*, Vol. 33, No. 7, July 1988, pp. 640-649.

[10] Hsu C.H.; Chen S.C.; Yu C.H. & Park J.H. (2009). Alleviating reader collision problem in mobile RFID networks, *Personal and Ubiquitous Computing*, Vol. 13, No. 7 October 2009, Springer-Verlag.

[11] Junius K.H. (2003). Solving the reader collision problem with a hierarchical q-learning algorithm. Master's thesis, Massachusetts Institute of Technology, February 2003.

[12] Kim, S.Y.; Lee J.K. (2009). A study on control method to reduce collisions and interferences between multiple RFID readers and RFID tag. *2009 International Conference on New Trends in Information and Service Science NISS*, 2009, pp.339-343.

[13] Kriplean, T.; Kumar, R.; French R.S. & Ramachandran, U.; (2007). Physical Access control for captured RFID Data. *IEEE Pervasive Computing*, Vol. 6, No. 4, 2007, pp.48-55.

[14] Liu D. (2009). ALOHA algorithm considering the slot duration difference in RFID system, *2009 IEEE International Conference on RFID*, 2009, pp. 56-63.

[15] Myug J. & Srivastava J. (2006). Adaptive Binary Splitting for Efficient RFID Anti-collision, *IEEE Communication Letters*,Vol. 10, No. 3, March 2006, pp. 144-146.

[16] Naware V.; Mergen G. & Tong L. (2005). Stability and delay of finite-user slotted ALOHA with multipacket reception, *IEEE Trans. on Info. Th.*, Vol. 51, No. 7, July 2005, pp. 2636-2656.

[17] Sabesan, S. ; Crisp, M. ; Penty, R.V. & White, I.H. (2008). Demonstration of improved passive UHF RFID coverage using optically-fed distributed multi-antenna system. *IEEE International Conference on RFID*, 2009, pp. 217 - 224.

[18] Samano-Robles; R. & Gameiro, A. (2009). Collision resolution algorithms for RFID applications. *Asia-Pacific Microwave Conference*, 2008, pp.1-4.

[19] Samano-Robles, R.; & Gameiro, A. (2009). Integration of RFID readers into wireless mobile telecommunications networks. *First Int. Conf, on wireless commun., vehicular tech., info. theory, and aerospace & electronics sys. tech., Wireless VITAE*, 2009, pp.327-331.

[20] Samano-Robles, R.; Ghogho M. & McLernon, D.C. (2009). Wireless Networks with Retransmission Diversity and Carrier-Sense Multiple Access. *IEEE Transactions on Signal processing*, Vol. 57, No. 9, pp. 3722-3726.

[21] Samano-Robles, R.; Ghogho M. & McLernon, D.C. (2009). A multi-access protocol assisted by retransmission diversity and multipacket reception, *IEEE ICASSP*, Las Vegas, Nevada, 2008, pp. 3005-3008.

[22] Samano-Robles, R. & Gameiro, A. (2012). Joint design of RFID reader and tag anti-collision algorithm: A cross-layer approach, *Proc IARIA International Conf. on Digital Communications - ICDT*, Chamonix, France, Vol. 1, pp. 1 - 6, April, 2012.

[23] Shakkotari S.; Rappaport T.S. & Karisson P.C. (2003). Cross-layer design for wireless networks, *IEEE Communications Magazine*, Vol. 41, No. 10, October 2003, pp. 74-80.

[24] Srivastaya, V. & Montani, M.; (2005). Cross-layer design: a survey and the road ahead. *IEEE Communications Magazine*, Vol. 43, No. 12, December 2005, pp.112-119.

[25] Tong L.; Naware V.& Venkitasubramaniam P. (2004). Signal Processing in Random Access,*IEEE Signal Processing Magazine, Special issue on Signal processing for networking: An integrated approach* , vol. 21, no. 5, Sep. 2004

[26] Tsatsanis M.K.; Zhang R. & Banerjee S. (2000). Network-Assisted Diversity for Random Access Wireless Networks, *IEEE Trans. on Sig. Proc.*, Vol. 48, No. 3, March 2000, pp. 702-711.

[27] Wagner J.; Fischer R. & Günther W.A. (2007). The influence of metal environment on the performance of UHF smart labels in theory, experimental series and practice. *First Annual RFID Eurasia conference* 2007, pp. 1-6.

[28] Waldrop N.; Engels D.W.& Sarma E. (2003). An anticollision algorithm for the reader collision problem, *IEEE Conf. On Commun. (ICC '03)*, Ottawa, Canada, 2003, pp. 1206-1210.

[29] Weinstein R. (2005). RFID: A Technical overview and its application to the enterprise,*IT professional*, Vol. 7, No. 3, 2005, pp. 27-33.

[30] Xue Y.; Sun H. & Zhu Z. (2009). RFID Dynamic Grouping Anti-collision Algorithm Based on FCM, *Int. Joint Conf. on Bioinfo., Sys. Biol. and Intelligent Comp.,*2009,pp. 619-622.

[31] Yan X. & Zhu G. (2009). An Enhanced Query Tree Protocol for RFID Tag Collision Resolution with Progressive Population Estimation, *International Conference on Mobile Adhoc and Sensor Systems (MASS)*, 2009, pp. 935-940.

Chapter 2

Choosing the Right RFID-Based Architectural Pattern

Michel Simatic

Additional information is available at the end of the chapter

1. Introduction

RFID provides a way to connect the real world to the virtual world. An RFID tag can link a physical entity like a location, an object, a plant, an animal, or a human being to its avatar which belongs to a global information system. For instance, let's consider the case of an RFID tag attached to a tree. The tree is the physical entity. Its avatar can contain the type of the tree, the size of its trunk, and the list of actions a gardener took on it.

When designing an RFID-based application, a system architect must choose between three locations to store the information: a centralized database, a database locally attached to the device hold by each user of the application, or the tag itself. Each location leads to an RFID-based architectural pattern[1]. But how to choose the right architectural pattern? What are the application attributes which must be taken into account in order to make the right choice?

The state of the art does not bring satisfactory answers. Indeed, when an article describes a RFID-based architectural pattern, it does not mention the application attributes which lead to choose this architectural pattern. On the other hand, some books or articles present the qualities of architectural patterns. But they do not take into account specificities of RFID. For instance, EPCglobal provides a standardized answer [2]: the centralized architectural pattern. A mobile device, NFC-enabled for example, reads an identifier on the RFID tag, then sends a message to a server which associates the identifier to the avatar stored in a central database. Thanks to its simplicity, this architectural pattern is used by several applications. But, it requires a global computer network: Such requirement increases operational costs. Moreover, it does not withstand an important number of simultaneous RFID read operations. Thus this pattern does not fit all RFID-based applications. [3] presents the stakes of introducing RFIDs inside an enterprise. But it does not contain any system architecture

---

1 . An. architectural. pattern. is. a. description. of. element. and. relation. types. together. with. a. set. of. constraints. on. how. they. may. be. used. [1].

thoughts. In a survey about RFID in pervasive computing, [4] presents several application examples. Depending on the application, avatars are stored either in a central database or in the tags themselves. But the authors do not give any clues on why an application has chosen to store its avatars in a given location. On the other hand, [1] lists the attributes which must be accommodated in a system architecture. Above all, there are the functionalities which are required from the system. Then, orthogonal to these functionality attributes, there are quality attributes. The authors distinguish system quality attributes (availability, modifiability, performance, scalability, security, testability, and usability), business qualities (time to market, cost and benefits, and projected lifetime), and qualities about the architecture itself (e.g. conceptual integrity). But the authors do not focus on RFID specific features.

So we have analyzed several existing industrial or experimental RFID-based applications. Moreover, we have developed RFID-based applications. From this experience, we identify the relevant attributes to compare RFID-based architectural patterns. We present them in section 2. With these identified attributes and their different aspects, we analyze four RFID-based architectural patterns, used by applications to access the avatar of a tagged entity. In the *centralized architectural pattern*, the mobile device reads an identifier on the RFID tag; then it contacts a server which associates this identifier to the avatar stored in a central database or in a database distributed between several companies [2]. With the *semi-distributed architectural pattern*, each mobile device holds a local copy of a central database associating RFID identifiers to avatars [5]. In the *distributed architectural pattern*, each RFID tag holds the avatar [6]. With the *RFID-based Distributed Shared Memory*, RFID tags hold the avatar and a replica of the avatar of other tags [7]. Sections 3 to 6 detail all of these architectural patterns: they present application examples and analyze the architectural pattern with the attributes identified in section 2. Thanks to this analysis, in section 7, we are able to provide guidelines to choose the convenient RFID-based architectural pattern. Finally, section 8 concludes this chapter and proposes perspectives for this work.

2. Architecture attributes and RFID technology

Relying on the experience gained by analyzing existing RFID-based applications and by developing RFID-based applications, we outline three architecture attributes among the attributes presented in [1]: (i) functionality, (ii) scalability, and (iii) cost. For each attribute, we present its different aspects which are influenced by the use of RFID technology.

2.1. Functionality attribute

Functionality is the ability of the system to do the work for which it was intended.

All architectural patterns give the ability to read/write the avatar of a read tag.

A first aspect of the functionality attribute is to check how the application behaves when it queries the avatar of a read tag. Is it guaranteed that the returned avatar has indeed the value which was last written? In other words, is there a staleness issue of avatar of a read tag?

The second aspect concerns the possibility of knowing the value (or having an order of idea of the value) of the avatar of a remote tag. By "remote", we mean that the user is not physically near the tag: The user is not able to put her reader on the tag. All she has is the identifier of the remote tag.

The third aspect is the staleness issue of the avatar of a remote tag. If the user is able to know the avatar of a remote tag, is it guaranteed that the returned avatar has indeed the last values associated to the tag?

2.2. Scalability attribute

The scalability criteria category evaluates how each architectural pattern behaves when there are numerous tags or numerous readers.

Its first aspect is the maximum number of tags which can be handled by the architectural pattern.

The second aspect characterizes the sensitivity of the architectural pattern to the number of simultaneous RFID tag read operations.

2.3. Cost attribute

The cost attribute groups all of the aspects which have an influence on the installation costs or the operational costs of the RFID-based application.

The first cost aspect concerns the requirement for a global network: do RFID readers have to be able to access at any time and any place to a specific computing machine (for instance, a server in the case of the centralized architectural pattern)? To fulfill this requirement, the readers may be equipped with a wired connection. In that case, the mobility of the readers is limited. The readers may also rely on Bluetooth® or Wi-Fi gateways. Both of these gateways may introduce installation costs. Moreover some readers may not be Wi-Fi enabled. For instance, the Nokia 6212 mobile phone is NFC-enabled, but has no Wi-Fi capabilities. Finally the reader may rely on a mobile data connection (e.g. UMTS, HSDPA, etc.). Such solution introduces operational costs because of data plans.

The second cost aspect concerns the RAM requirement on each tag. The more RAM there is on the tag, the more expensive the tag is. Notice that RAM may actually be prohibited on tags for technical reasons and not for cost reasons. For instance, application may require the use of low-frequency tags (e.g. 125 kHz), so that readers can interact with tags even though there is a liquid between tags and readers. In this case, the throughput is too low for a tag to host information other than its identifier.

The third cost aspect concerns the introduction of a new tag in the environment. For each architectural pattern, we determine the sequence of operations which is required in order to introduce a new tag in the environment. Knowing this sequence, we can determine how long this sequence lasts. Because this initialization procedure is executed by a human or a robot operator, its cost is proportional to the time spent.

The final cost aspect is related to the reinitialization of all of the tags. This criterion concerns only applications which, during their lifetime, need sometimes to have each tag given a new initial value. For instance, this is the case of Paris public transportation system. Users are equipped with a transportation pass containing an RFID tag. At the beginning of a month, each user has to reload her pass (to refresh her access rights): in other words, the tag has to be reinitialized. Some RFID-based games also require tag reinitialization. Indeed, in the case of non-permanent games, users play during successive game sessions. Thus at the beginning of each session, all of the tags must be reinitialized.

In this section, we have defined different aspects of three architecture attributes: (i) functionality, (ii) scalability, and (iii) cost. These aspects are influenced by the use of RFID technology. We use them to compare the behavior of four RFID-based architectural patterns. We start by analyzing centralized architectural pattern.

3. Centralized architectural pattern

This architectural pattern is often used by manufacturing applications. It has been standardized by EPCglobal [2]. When a reader is near a tag (for instance, the blue mobile in Figure 1), it reads the tag's identifier or an identifier stored in the tag's data zone (its Electronic Product Code in the case of EPCglobal). This identifier is represented by the hexagon in Figure 1. Then, the reader asks a server (ONS lookup service in the case of EPCglobal) which machine (EPC Manager in the case of EPCglobal) manages the avatar corresponding to the read identifier. When the server responds, the reader contacts this machine with the identifier of the tag. The machine queries its database and returns the avatar (for instance, the contents of the hexagon in the database in Figure 1).

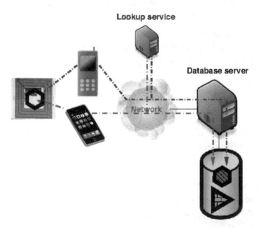

Figure 1. Centralized architectural pattern

Next section gives examples of this architectural pattern.

3.1. Examples

Aspire RFID is an Open Source middleware which is compliant to the specifications of EPC-global [8]. It proposes several examples of industrial applications for tracking goods.

Next paragraphs present products or prototypes developed according to centralized architectural pattern, but without being compliant to EPCglobal.

PAC-LAN is a game prototype in which players are equipped with NFC mobile phones without any GPS capabilities [9]. Players must interact with NFC tags which have been disseminated throughout a neighborhood. In a central database, the identifier of each tag is associated to geographical coordinates. When a player reads a tag, her mobile phone uses the UMTS network to contact the server with the tag's identifier. The server queries its database, finds the geographical coordinates, and broadcasts them to all of the players. An administrative application is provided to reset a game on the server. Such reset has an impact on all of the players' mobile phones.

[10] proposes an application so that visitors of an art exhibition can discover the paintings in another way. NFC tags are dispatched on the back of exposed paintings. Equipped with an NFC-enabled phone, the visitor puts her phone on spots of the paintings which intrigue her. Phone reads the identifier stored in the tag. Then, it contacts an uGASP server [11-12]. After consulting an internal database, this server indicates to the mobile phone what must be done: display a text, an image, or play an audio comment. Thus the author of the painting is able to communicate with the visitor.

Via Mineralia is a pervasive serious game which goal is to enrich the visit of a Freiberg museum [13]. In this game, the visitor uses a PDA equipped with an RFID reader. RFID tags (holding a unique identifier) are dispatched in the showcases which the museum wants to emphasize. When the PDA scans a tag, it sends an HTTP request (with tag's identifier) to a web server. To do so, the PDA uses a Wi-Fi network which covers the whole museum. The server answers to the PDA with multimedia information. The PDA displays them in a navigator.

Touchatag company (formerly Tikitag) sells NFC readers which can be connected to Windows or Mac-OS personal computers, and NFC tags dedicated to Touchatag [14]. A customer can then connect to http://www.touchatag.com web site, and define the reaction to be associated to the reading of one tag. When the NFC reader reads a tag, it contacts the Touchatag application which runs permanently on customer's personal computer. Then, via the Internet network to which the computer is connected, this application contacts a Touchatag service called *Application Correlation Service* (ACS). Touchatag application gives tag's identifier to ACS. Then, ACS queries Touchatag database to find reaction associated to the reading of this tag. It sends back this information to Touchatag application. The touchatag application reacts in the appropriate way. For instance, let's assume that the customer has specified the following action on Touchatag web site: when tag r with identifier i is put on the reader, customer wants her browser to access to Uniform Resource Locator (URL) of a web site w.

Then, when customer puts tag r on the reader, Touchatag application contacts ACS with identifier i. ACS replies with URL of w. Then, Touchatag application opens a browser with this URL w.

Skylanders is a video game developed by *Activision* company [15]. It requires the use of plastic figures. These figures contain an NFC tag. When a player puts her figure on top of a "Portal of Power" (actually, an NFC tag reader), the video game reads the identifier stored in the NFC tag. Then, the game contacts a server to get the information concerning the character which must be displayed: The figure becomes alive on the screen. Notice that, according to [16], information is also stored inside the tag: Thus the game can work without using a global network to contact a server. This means that Skylanders not only uses a centralized architectural pattern, but also a distributed one.

Based on all of these examples, next section analyzes centralized architectural pattern.

3.2. Analysis

Concerning the functional attribute, any transponder which wants to modify the avatar of a tag does so by sending a modification message to the server. Thus the server is always aware of the last update done on any avatar. As a reader always queries the central database to know the avatar of a tag, it is not possible that the read value is stale. Moreover, knowing a tag identifier, a reader is able to query the server to know the avatar associated to this identifier: the reader is able to know the avatar of a remote tag. As a mobile device queries the server to know the avatar of a remote tag, it is sure that the returned value is not stale.

Concerning the scalability attribute, the maximum number of tags which can be handled by this architectural pattern is limited by the number of avatars which can be stored in the central database. Let s be the average size in bytes of an avatar. Let $S_{central}$ be the maximum size in bytes of the database. We neglect the storage of the link between tag identifiers and avatars in the database. Moreover, we neglect the overhead due to the storage of data in the database. Then, the maximum number of tags is bounded by $S_{central}/s$. About sensitivity to the number of simultaneous reads, this architectural pattern is restrained by its centralized nature. The server holding the ONS lookup service may become a bottleneck. Moreover, the different servers of avatars may not return avatar values fast enough. Of course, it is possible to increase the number of servers. But that makes the hardware architecture more complex and more costly (from an installation and a management point of view). Thus this architectural pattern may not be applicable for some applications.

Concerning the cost attribute, the reader must always be in contact with the server holding the ONS lookup service and the servers of avatars: a global network is required. On the other hand, this architectural pattern only needs to read an identifier on the tag. And this identifier can be stored in ROM as it is never modified: no RAM is required on the tags. When a new tag is introduced in the system, three operations are required: (i) the tag is linked to the physical entity; (ii) the avatar of this entity is initialized in the central database; and (iii) a link between the tag identifier and this avatar is created into the central database. When all of the tags have to be reinitialized, a program is run on the server hosting the central database. It sets each avatar to its new value.

This section has analyzed centralized architectural pattern according to the attributes presented in section 2. This architectural pattern fulfills all aspects of functionality attribute. But this is achieved with the operational cost of a global network. Another disadvantage is a high sensitivity to the number of simultaneous read operations.

Next section analyzes semi-distributed architectural pattern which compensates the requirement for a global computer network and reduces sensitivity to the number of simultaneous reads.

4. Semi-distributed architectural pattern

In semi-distributed architectural pattern, mobile RFID-enabled devices (PDAs, mobile phones, etc.) are periodically synchronized with a central database holding all of the avatars (see Figure 2). Then, human operators carry the mobile devices near the entities to which the tags are associated. When a device comes close to an entity, the device reads the identifier of the entity's tag. By querying its local copy of the central database, the device is able to find the avatar of this entity. Any modification of an avatar is done on the local copy. It is propagated to the central database at the next synchronization.

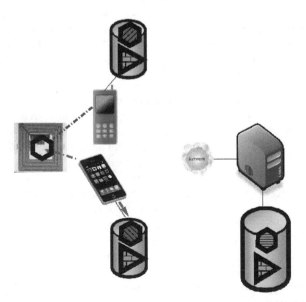

Figure 2. Semi-distributed architectural pattern

Next section presents an example of this architectural pattern.

4.1. Example

The unique example of use of such architectural pattern is Paris trees management application [5]. Each of the ninety-five thousand trees of Paris avenues is equipped with an RFID tag. Each gardener synchronizes her tablet PC with the central database before a new day of work. During her day of work, whenever a gardener does something to a tree, she identifies the tree thanks to its RFID tag: Her tablet PC modifies the avatar in the local database. Then, in the evening, she synchronizes her tablet PC with the central database. Thus she uploads her database updates and downloads updates from other gardeners.

Now, let's analyze the semi-distributed architectural pattern.

4.2. Analysis

Concerning the functional attribute, the avatar of a read tag may be stale. Suppose users U_1 and U_2 synchronize their mobile device with the central database. Then U_1 modifies avatar of tag r. Thus she modifies her local copy of the database. When U_2 comes to tag r, as her device reads its local copy of the database, the returned value of the avatar is the value before U_1's modification: the read value is stale. Notice we can limit this issue by assigning sets of entities to each mobile device. For instance, in the case of the Paris trees management application, a supervisor can assign a set of trees to be taken care of during the day, to each of the gardeners. If all of these sets are apart, this issue cannot be observed anymore. About remote tags, by querying its local database, the device is able to read the avatar of a remote tag, even though there is no global network. But the read value can be stale. It will be again correct only when all of the mobile devices have synchronized themselves with the central database.

Concerning the scalability attribute, the maximum number of tags which can be handled by this architectural pattern is limited by the number of avatars which can be stored in the central database and in the local copy of this database. Let S_{local} be the maximum size in bytes of the local database. The maximum number of tags is bounded by $min(S_{central}/s, S_{local}/s)$, which is likely to be S_{local}/s as mobile devices do not have as much memory as servers. Notice that this bound can be increased to $S_{central}/s$ by assigning to each mobile only a subset of the central database. For instance, in the case of Paris trees management application, the mobile device of a gardener could receive only the avatars of the trees she will take care of during the day. About sensitivity to the number of simultaneous reads, this architectural pattern is not as sensitive as centralized architectural pattern. It does not need to query a server upon each RFID tag read. Nevertheless all of the readers must periodically synchronize themselves with the central database. As the synchronization time is proportional to the number of readers, it may reach unbearable values. This issue can be tackled by limiting the number of avatars copied on the local devices, thus reducing the volume of data transferred between each device and the central database.

Concerning the cost attribute, the mobile RFID-enabled devices only need an access to the server hosting the central database during synchronization phase. At that moment, devices are probably near the central database: A Wi-Fi network may be used. Otherwise it is the local database which is queried. Thus no global communication network is required around the working area. Moreover, as in centralized architectural pattern, there is no need for RAM on the tags. When a new tag is inserted in the system, the procedure to be applied is the same as in the centralized architectural pattern. When all of the tags have to be reinitialized, a program is run on the server hosting the central database. It sets each avatar to its new value. However, the reinitialization of the tags will be effective only when all mobile devices will get synchronized with the central database.

This section has analyzed semi-distributed architectural pattern according to the attributes presented in section 2. This architectural pattern does not require any global network and has a medium sensitivity to the number of simultaneous tag reads. Nevertheless it faces a functional issue concerning the staleness of avatar read on a local (or remote) tag.

Next section analyzes the distributed architectural pattern which tackles the sensitivity and staleness issues.

5. Distributed architectural pattern

In distributed architectural pattern, the avatar of an RFID tag is stored inside the RAM of the tag (see Figure 3). Whenever a user is in contact with a tag, the reader works with the part of the RAM containing the avatar.

Figure 3. Distributed architectural pattern

Next section gives examples of this architectural pattern.

5.1. Examples

Nokia 6131 NFC phones are sold with three NFC tags. Each one triggers a different function on the telephone: One activates alarm function; another one plays a given music on the phone; the last one displays an NFC tutorial. To do so, the telephone reads the contents of the tag, this contents being coded as a Uniform Resource Identifier (URI) according to the NFC Forum's specifications of *Smarts Posters* [5,17-18]. When the phone is programmed to understand tags' contents formatted according to these specifications, these URIs can be used to tell the telephone to accomplish a given function like send an SMS, call a certain number, open a given web page, etc.

In fact, it is thanks to this Smart Posters specification that any NFC phone can exploit Touchatag tags mentioned in section 3.1. Indeed, these tags contain not only an identifier used by Touchatag application, but also an URI. This URI is the URL of a Touchatag web server with a parameter containing the identifier of the tag. Thus when a user touches a Touchatag tag with her mobile phone, the phone reads the URL and then opens a browser with this URL. Touchatag web server is then contacted, via 3G or Wi-Fi, with the identifier i. Then, web server contacts ACS (see section 3.1) with i. In the case where i is associated with a web site w, URL of w is sent back to Touchatag web server. This server returns an html page containing a redirection towards w. Finally, the browser displays w. Notice that, in the case of a Touchatag tag read by a mobile phone, phone uses distributed architectural pattern to determine the Touchatag web server to contact; but, the Touchatag web server uses centralized architectural pattern to translate the tag identifier into an action.

Once again, it is the Smart Posters specification which is used by *Connecthing* company to bring intelligence to mailboxes [19]. When a user scans a mailbox equipped with an NFC tag, her phone reads the URL stored on the tag (which contains an identifier corresponding to the physical location of the mailbox) and opens a browser to access this URL. This web page displays location of nearby mailboxes, the time at which postman takes the mail, etc.

Navigo, the Paris public transportation pass, is an example of an industrial application based on this architectural pattern, which does not use Smart Posters specification [20]. The 4.5 million Navigo pass users do not have an NFC reader. They are only given a pass which contains an NFC tag. With a vending machine, each user initializes her tag with the rights she buys to use the public transportation. Whenever she wants to use a public transportation, she presents her pass in front of an NFC reader. Locally, the reader checks the rights stored in the tag's RAM and opens the gate, if the access is granted.

Ubi-Check is an academic application example of distributed architectural pattern [21]. An RFID tag is attached to each of a traveler's items. At the beginning of their travel, each tag is initialized with a value specific to the traveler. All of these RFID tags are read after special points (e.g. after an airport security control). If an inconsistency is found among the read values, it means that, at some point, the traveler exchanged one of her items with the item of another traveler. An alarm is thus triggered to warn the traveler that one of her items is missing.

[22] proposes an academic system based on digital pheromones to find objects lost in a house. To do so, floor of the house is covered with RFID tags. An RFID reader is coupled with each house object. When user moves an object from point *A* to point *B*, the RFID reader associated with the object behaves like an ant which sets pheromones on the path it takes: The reader writes a digital pheromone (made of object identifier and timestamp of transit) in the RAM of each tag over which it goes. Notice that, like a natural pheromone which evapo-rates with time, whenever a reader finds no more room in the RAM of a tag (there are too many pheromones stored inside), the reader deletes the oldest pheromone from the tag. In case an object is lost, user takes a dedicated RFID reader and wanders around the house un-til she finds the digital pheromones of the object. Once she has located it, she follows the pheromone trace until the place where the object was left.

Roboswarn is an (academic) application to position robots (equipped with NFC readers) in a physical space to accomplish a certain task [23]. NFC tags are dispatched in dedicated places of a room (for instance, near a hospital bed which these robots will have to push so that a cleaning robot can accomplish its task). Each tag is initialized with location of other tags in the room and the timestamp of last cleaning. When robots enter the room, they look for an NFC tag. As soon as one robot finds one, it reads the position of other tags and transmits them to other robots. The other robots go to the other tags. If timestamp of last cleaning is too old, robots push the hospital bed and then write new timestamp of cleaning. Otherwise, robots do nothing.

SALTO Systems company is selling locks for electronic doors. The keys are NFC tags. To fa-cilitate the management of all locks and tags, this company has developed SALTO Virtual Network (SVN) [24]. Thanks to this system, Heathrow airport operator is able to manage 1000 standard electronic locks (NFC-controlled) and 37 hot spots. These spots are special locks connected to a global computer network. They can: 1) unlock an entry access on the whole site, 2) initialize an NFC key with the right to open given locks during the day, 3) blacklist some NFC keys, 4) recover data collected by the key during the working day of its user. Indeed, each time a person unlocks an electronic lock with her NFC key, the lock reads data stored on tag to check user permissions and the list of blacklisted tags. But, the elec-tronic lock also writes information like, for instance, the low charge of the battery powering the lock. Thus thanks to SVN, even though standard locks do not have access to a global computer network, they can receive information (e.g. list of blacklisted cards) and send in-formation (e.g. low charge of battery): Standard locks communicate thanks to the network made of the users of the keys/tags.

Based on all of these examples, next section analyzes distributed architectural pattern.

5.2. Analysis

Concerning the functional attribute, as the avatar is written and read only in the RAM of the tag, there is no staleness issue of locally read tags. However, it is impossible to know the avatar of a remote tag.

Concerning the scalability attribute, there is no limit on the number of tags in the application environment. Moreover, such distributed architectural pattern is not sensitive at all to the number of simultaneous read operations (all of the operations are done locally).

Concerning the cost attribute, the reader does not need any global network to access to the avatar of the RFID tag. On the other hand, RAM is required on each tag. Its size must be at least the size of the avatar. This means that the avatar cannot contain too much information (e.g. MIFARE tags can offer up to 4 Kbytes of RAM, with 3440 bytes of net storage capacity). When a new tag is introduced in the system, only two operations are required: (i) the tag is linked to the physical entity; and (ii) the avatar of this entity is initialized in the RAM. About the reinitialization of the tags, it is application-dependant. Some applications require that a dedicated user goes through all of the tags to reinitialize them. In the case of Navigo pass, users are in charge of bringing their pass to a vending machine. This leads to long waiting lines at the beginning of a month, when users must initialize their rights for this month. This is why Navigo operator carries out experiments where users can initialize their tag using a dedicated NFC reader connected to their personal computer. To avoid reinitialization costs, some RFID-based distributed applications put in place special mechanisms. These mechanisms take into account elapsed time in order to automatically reset data. In Roboswarm application (see section 5.1), there is no need to reset the timestamp to trigger a new cleaning of a room. Each robot is aware of a deterioration level. Thus if the timestamp plus this deterioration level is greater than current time, it means that the room needs some cleaning again. With application for pheromone-based object tracking (see section 5.1), although tags have limited RAM capabilities, there is still no need to have a periodic session initialization which would clean up outdated pheromones. Each pheromone is written on a tag with a timestamp. Thus when the device attached to the roaming object meets a tag, it cleans up pheromones which have a too old timestamp, before writing the dedicated pheromone.

This section has analyzed distributed architectural pattern according to the attributes presented in section 2. This architectural pattern does not require any global computer network. And it is not sensitive to the number of simultaneous read operations. Nevertheless it faces a functional issue: it is not possible to get the avatar of a remote tag.

Next section analyzes RFID-based DSM which tackles this issue.

6. RFID-based distributed shared memory architectural pattern

RFID-based distributed shared memory (RFID-based DSM) mixes the qualities of semi-distributed architectural pattern and distributed architectural pattern [7]. The avatar of an RFID tag is stored in the RAM of the tag. In addition, each tag and each mobile device of the application environment holds a local copy of all of the avatars (see Figure 4). Moreover they hold a vector clock (see Figure 5). Each element of this vector clock is a number corresponding to the last version of the avatar which the tag or the device has learnt about (this is why [25] gives the name *version number* to this number). When a mobile device comes to a tag and modifies the avatar of the tag, this mobile increments the element of the vector clock (stored on the tag and inside its own memory) corresponding to the avatar of this tag. Whenever a mobile device meets a tag (respectively another device), the device and the tag (respectively the other device) compare their respective view of the avatars, by comparing their vector clocks values. Doing so, each of them learns from the other one the latest news (which they are aware of) about all of the avatars.

Figure 4. RFID-based distributed shared memory architectural pattern

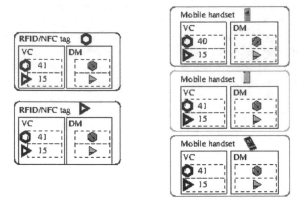

Figure 5. Data of RFID-based distributed shared memory

Next section presents an example of this architectural pattern.

6.1. Example

Plug: Secrets of the museum, an (academic) pervasive game [26] developed in the context of the *PLUG* research project [27], is the unique example of use of RFID-based DSM architectural pattern. In this game, 48 virtual playing cards represent objects of French Museum of Arts and Crafts (*Musée des arts et métiers*). These cards are dealt between 16 NFC tags (1 card per MIFARE tag, each of them being equipped with 1 KB of RAM) and 8 mobile phones (4 cards per Nokia 6131 NFC mobile). The players' goal is to collect cards of the same family on her mobile. To do so, players use their mobile to swap a card with a tag or another mobile.

Next section analyzes RFID-based DSM architectural pattern.

6.2. Analysis

Concerning the functional attribute, whenever a mobile device comes near a tag, there are two possibilities. Either the tag has been already initialized; in that case, as the avatar is stored on the tag, the value read on the tag is the most up-to-date. Or, the tag has not been already initialized; in that case, the first task of the mobile device is to initialize the tag; so that the value read on the tag after this initialization is also the most up-to-date value. Thus there is no staleness issue for avatar of locally read tag. Moreover, a mobile device holds a local copy of all avatars. Thus by querying this local copy, the device is able to answer to queries concerning a remote tag. However this local copy may not be up-to-date: There is a staleness issue for avatar of remotely read tag.

Concerning the scalability attribute, the maximum number of tags is limited by the size of the RAM of the tags. This architectural pattern stores copies of the avatar of all of tags and a vector clock. Let S_{tag} be the lowest size of the RAM of the tags present in the environment. Let L be the length of an element of the vector clock. Then the maximum number of tags is bounded by $S_{tag}/(s + L)$. Let's compare this bound to the bound of semi-distributed architectural pattern. For the latter pattern, the numerator is expressed in terms of Gigabytes. However it is expressed in terms of Kilobytes in the case of a tag's RAM. The maximum number of tags for RFID-based DSM architectural pattern is at least one million times lower than the maximum number for semi-distributed architectural pattern. About sensitivity to the number of simultaneous reads, RFID-based DSM requires that all mobile devices synchronize with a dedicated machine: RFID-based DSM is as sensitive as semi-distributed architectural pattern.

Concerning the cost attribute, in RFID-based DSM architectural pattern, the reader does not need any global network to access to the avatar of the RFID tag. Nevertheless RAM is required on each tag. Its size must be at least the size of the avatar times the number of tags in the environment (so that a tag can hold a local copy of all avatars). This means that the avatar can contain even less information than in the case of distributed architectural pattern.

When a new tag is introduced in the system, four operations are required: (i) the tag is linked to the physical entity; (ii) the avatar of this entity is created and initialized in DB_{init}, the database used to (re)initialize the local copy on each mobile (DB_{init} is stored on a dedicated machine which can be one of the mobiles); (iii) a link between the tag identifier and this avatar is created in DB_{init}; and (iv) all of the elements of the tag's vector clock are initialized to zero.

When tags must be reinitialized, a program P_{init} is executed on the machine hosting DB_{init}. This program computes the initial value VC_{init} of the vector clock for this session, so that each element of VC_{init} is greater, thus more recent, than all the vector clock elements in the mobiles. To do so, P_{init} can use two methods. The first method is twofold: (i) get the vector clocks of all of the mobiles; and (ii) compute the maximum value. This first method does not require additional memory on each tag, but requires additional communication between the mobiles and the dedicated machine. This method works because the vector clock of a tag evolves only when a tag is in contact of a mobile. Thus there is always at least one mobile device which is aware of the values stored in the vector clock of a tag: P_{init} does not need to be aware of the vector clock values stored on the tags. The second method supposes that each vector clock element is made of two fields: a "session identifier" field and a "tick in this session" field. Thus P_{init} has only to increase the session number and set all "tick in this session" fields to zero. This second method requires additional memory on each tag, but no additional communication between the mobiles and the machine running P_{init}. The choice of the method is application dependant. Once one of the two methods has been applied, the dedicated machine synchronizes each mobile device by sending the contents of DB_{init} and VC_{init} to the device. Afterwards, whenever a mobile device is in contact with an uninitialized RFID tag for this session, as the mobile device vector clock is greater than the tag vector clock, the mobile device initializes the tag. In other words, RFID-based DSM architectural pattern takes advantage of the fact that application users will go to tags, to initialize them: This pattern uses the communication network made by application users, instead of using a global computer network.

This section has analyzed RFID-based DSM architectural pattern according to the attributes presented in section 2. This architectural pattern does not require any global computer network. And it does not experience the issue of staleness of an avatar of a read tag. Moreover it is possible to query the avatar of a remote tag. Nevertheless this architectural pattern experiences staleness issues when accessing to avatar of a remote tag. And there is a scalability issue in terms of maximum number of tags which can be handled.

By synthesizing the conclusions observed for the different architectural patterns, the next section provides guidelines for choosing the most adequate pattern for a given application.

7. Guidelines for choosing the right RFID-based architectural pattern

Table 1 synthesizes the analysis of the different aspects of the architecture attributes made on all of the architectural patterns. In this table, values which are in italic correspond to aspects which are a limitation for this architectural pattern.

If the application requires the best level for all aspects of functionality attribute, then the centralized architectural pattern must be chosen. It is the only architectural pattern which experiences no issues within the functionality attribute. But this pattern has an operational cost due to the requirement for a global network. And this pattern is highly sensitive to the number of simultaneous reads.

	Central.	Semi-distr.	Distributed	RDSM
Staleness of locally read tag	No	*Yes*	No	No
Avatar of remote tag	Yes	Yes	*No*	Yes
Staleness of remote read tag	No	*Yes*	n.a.	*Yes*
Maximum number of tags	$S_{central}/s$	S_{local}/s	Infinite	$S_{tag}/(s+L)$
Sensitivity to number of simultaneous reads	*High*	Medium	None	Medium
Network required	Yes	No	No	No
RAM required on tag	No	No	*Yes*	*Yes*
Cost of introducing a tag (most costly operation)	Link tag to physical entity	Link tag to physical entity	Link tag to physical entity	Link tag to physical entity
Cost of reinit. Tags (most costly operation)	Reinit. Database	Sync. Database	*Go to all tags*	Sync. database

Table 1. Comparison of the RFID-based architectural patterns (italic values signal a limit for this architectural pattern)

If one of these last two issues is a problem, the system architect must consider the three other architectural patterns. Semi-distributed architectural pattern must be chosen if RFID tags cannot host RAM. This constraint may be due to cost motivations, but also technical constraints (use of low-frequency tags, see section 2).

If there can be RAM on tags, the maximum number of tags must be determined. If it is compatible with RFID-based DSM architectural pattern limitations, then this pattern should be chosen (as it is the least limited pattern for the functionality attribute). Otherwise the system architect should choose distributed architectural pattern (if there must be no staleness issue for read tags) or semi-distributed architectural pattern (if the cost of reinitializing tags is an important factor). Notice that the mixing of distributed architectural pattern and RFID-based DSM may be an interesting alternative. On each tag, we can store its avatar and the vector clock element corresponding to this avatar. Each mobile device holds a copy of all avatars and a full vector clock. By applying RFID-based DSM procedures, we get a solution for the limited maximum number of tags in RFID-based DSM. And in the same time, we solve distributed architectural pattern limitations (as we can query avatar of remote tags and we reduce the high cost of tag reinitialization).

To illustrate the use of these guidelines, let's consider the choice of the architectural pattern for the RFID-based game *Plug: Secrets of the museum* presented in section 6.1.

Each tag costs about 0.10 euro (respectively 1.50 euro) if it has 0 KB (respectively 1 KB with 752 bytes of net storage capacity, S_{tag}=752 bytes) of RAM. The avatar of a tag is the virtual card "contained" in the tag. There is a maximum of 16 cards in the game. Thus the avatar is coded as a byte value (s=1 byte). Concerning the vector clock, the project uses the synchronization method requiring a session identifier. To have an ever-increasing value, "session identifier" field stores the initialization time. This time is the difference, measured in milliseconds, between the session initialization time and midnight, January 1st, 1970 UTC. This storage requires 8 bytes per tag. Moreover, each tag holds the "tick in this session" field of each avatar stored in the tag.

This field is coded as a short (L=2 bytes). It represents a real-time clock, formatted as the number of seconds since the beginning of the game session (A session lasts less than 2 hours: there is no risk of overflow). This clock is the time known by the tag of the last update of the avatar of another tag. It takes about 20 minutes to attach each of the 16 tags to their correct location, so an average of 75 seconds per tag. Linking the tag and the avatar takes about 5 seconds per tag. Initializing the avatar is done in a few milliseconds by an initialization program. For reinitializing tags, synchronizing all of the 8 mobiles with a dedicated machine takes about 1 minute, that is an average of 4 seconds per tag. Notice that synchronization is based on NFC peer-to-peer communication. The project could have saved synchronization time if it has used Bluetooth®, but it would have used more battery.

If the project is going to use centralized or semi-centralized architectural pattern, it will use a dedicated machine for hosting the central database. This machine will be equipped with a 500 gigabytes disk ($S_{central}$=500 GB). In the case of the semi-distributed architectural pattern, the project will use half of the micro-SD memory of each mobile phone to host the local copy of the database (S_{local}=1 GB). If the project is going to use centralized architectural pattern, each mobile will have to be equipped with a SIM card giving access to a UMTS data plan. This will cost 15 euros per mobile per month. If the project is going to use distributed architectural pattern, it will take 13 minutes to go by all of the 16 tags to reinitialize them, so an average of 49 seconds per tag.

We apply these numeric values to table 1. Table 2 synthesizes the results. In this table, values which are in italic correspond to criteria which are a limitation for this architectural pattern.

Centralized architectural pattern requires a global network which costs 120 euros per month. The museum which hosts the game considers it is too expensive. We have to turn to one of the other architectural patterns. As the game must manage 16 tags and as the RFID-based DSM can handle a maximum of 248 tags (as s=1 byte), we can choose this architectural pattern. However, if s had been 250 bytes, this pattern could have handled only 4 tags: It would not have fitted. As there must be no issue about avatar of read tags (the game would not be fun), we would have chosen distributed architectural pattern (or combination of distributed and RFID-based DSM patterns, in order to reduce the costs of reinitializing tags).

8. Conclusion and future work

This chapter studies four RFID-based architectural patterns: centralized, semi-distributed, distributed and RFID-based DSM. It compares them according to nine aspects of three architecture attributes: functionality, scalability and cost. Despite their specific limitations, each architectural pattern fits the requirements of existing applications.

	Central.	Semi-distr.	Distributed	RDSM
Maximum number of tags if s=1 byte (s=250 bytes)	500 x 109 (2 x 109)	109 (4 x 106)	Infinite (Infinite)	248 (2)
Cost of computer network (per month)	*120 euros*	0 euro	0 euro	0 euro
Cost of tag (per tag)	0.10 euro	0.10 euro	*1.50 euro*	*1.50 euro*
Cost of introducing a tag (in seconds per tag)	80 s/tag	80 s/tag	80 s/tag	80 s/tag
Cost of reinitializing tags (in seconds per tag)	0 s/tag	4 s/tag	*49 s/tag*	4 s/tag

Table 2. Comparison of the RFID-based architectural patterns in the case of the game *Plug: Secrets of the Museum* (italic values signal a limit for this architectural pattern)

The chapter proposes guidelines for choosing the RFID-based architectural pattern which will best fit a given application requirements. These guidelines are tested in the context of an RFID-based pervasive game.

Future work concerns the analysis of these architectural patterns with respect to security architecture attribute. Security is a measure of the system's ability to resist unauthorized usage while still providing its services to legitimate users [1]. This future work will determine the influence which the level of resistance to security attacks and the cost of implementing such resistance have on the guidelines provided in this paper. Another attribute we would like to study is the fault-tolerance of the different elements of the system.

Author details

Michel Simatic

Address all correspondence to: Michel.Simatic@telecom-sudparis.eu

INF Department, Télécom Sud Paris, Évry, France

References

[1] Bass L., Clements P., and Kazman R., Software Architecture in Practice, 2nd Edition. Addison-Wesley Professional, April 2003, ISBN-13: 978-0-321-15495-8.

[2] Armenio F., Barthel H., Burstein L., Dietrich P., Duker J., Garrett J., Hogan B., Ryaboy O., Sarma S., Schmidt J., Suen K., Traub K., and Williams J., "The EPCglobal architecture framework," GS1 EPCglobal, Tech. Rep. Version 1.2, September 2007.

[3] Gonzalez L., RFID: Stakes for the enterprise! (in French). Afnor Editions, October 2008, ISBN-13 978-2-12-465153-5.

[4] Roussos G. and Kostakos V., "RFID in pervasive computing: State-of-the-art and outlook," Pervasive Mob. Comput., vol. 5, no. 1, pp. 110–131, February 2009.

[5] ITR Manager.com, "City of Paris is taking care of its trees with RFID tags (in French)," http://www.itrmanager.com/articles/59758/59758.html (last access in July 2012), December 2006.

[6] "Smart Poster Record Type Definition, Technical specification SPR 1.1," NFC Forum, 2006.

[7] Simatic M., "RFID-based replicated distributed memory for mobile applications," in Proceedings of the 1st International Conference on Mobile Computing, Applications, and Services (Mobicase 2009), San Diego, USA. ICST, October 2009.

[8] Aspire RFID: OW2 Aspire RFID: an RFID suite for SMEs. Available at http://wiki.aspire.ow2.org (last access in July 2012), 2011.

[9] Rashid O., Bamford W., Coulton P., Edwards R., and Scheible J., "PAC-LAN: mixed-reality gaming with RFID-enabled mobile phones," Computers in Entertainment, vol. 4, no. 4, pp. 4–20, October–December 2006.

[10] Haberman O., Pellerin R., Gressier-Soudan E. et Haberman U. (2009). RFID painting demonstration. In Natkin S. and Dupire J., éditeurs : Entertainment Computing - ICEC 2009, volume 5709 de Lecture Notes in Computer Science, pages 286-287. Springer Berlin / Heidelberg.

[11] Pellerin R., Adgeg G., Delpiano F., Gressier-Soudan E., and Simatic M., "Gasp : a middleware for mobile multiplayer games". http://gasp.ow2.org (last access in July 2012), July 2007.

[12] Pellerin R., Delpiano F., Duclos F., Gressier-Soudan E. and Simatic M. (2005) "Gasp : an open source gaming service middleware dedicated to multiplayer games for J2ME based mobile phones". In 7th Int. Conference on Computer Games CGAMES'05 Proceedings, pages 75-82.

[13] Heumer G., Gommlich F., Jung B. and Müller A. (2007). Via Mineralia - a pervasive museum exploration game. In Proc. of Pergames 2007, Salzburg, AT.

[14] Touchatag: Using the Advanced HTTP Application. Available at http://www.touchatag.com/developer/docs/applications/advanced-HTTP (last access in July 2012), 2010

[15] Activision: Skylanders : Spyro's adventure. Available at http://www.skylanders.com/) (last access in July 2012), 2011.

[16] Planck S.: Kids going crazy for Activision Skylanders NFC game. Available at http://www.nfcrumors.com/01-17-2012/kids-crazy-activision-skylanders-nfc/) (last access in July 2012), January 2012.

[17] NFC Forum. NFC Data Exchange Format (NDEF) - Technical Specification 1.0. Rapport technique, NFC Forum.

[18] NFC Forum (2006c). URI Record Type De?nition - Technical Specification 1.0. Rapport technique, NFC Forum.

[19] Connecthings : 08/12/2011 - BAL intelligente @ mobulles Paris @ Lab Postal 2011 (Intelligent mailbox, in French). http://www.connecthings.com/node/123 (last access in July 2012), December 2011.

[20] Levallois-Barth C., "Navigo: simplification or absolute traceability". FYP éditions, May 2009, ch. 5 – Security and data protection, in Evolution of digital cultures (in French), pp. 173–181, ISBN-13 978-2-91-657113-3.

[21] Couderc P. and Banâtre M., "Beyond RFID: The Ubiquitous Near-Field Distributed Memory," ERCIM news, no. 76, pp. 35–36, January 2009.

[22] Mamei M. and Zambonelli F. (2007). Pervasive pheromone-based interaction with RFID tags. ACM Trans. Auton. Adapt. Syst., 2(2):4.

[23] Zecca G., Couderc, P., Banâtre M. et Beraldi R. (2009). « Swarm robot synchronization using RFID tags. In Pervasive Computing and Communications," 2009. PerCom 2009. IEEE International Conference on, pages 1-4.

[24] SALTO Systems: SALTO Networked Locking System - SVN. Available at http://www.saltosystems.com/index.php?option=com_content&task=view&id=62&Itemid=57) (last access in July 2012), 2007.

[25] Murphy A. L. and Picco G., "Using LIME to Support Replication for Availability in Mobile Ad Hoc Networks," in Proceedings of the 8th International Conference on

Coordination Models and Languages (COORD06), vol. 4038, Bologna, Italy. Springer Lecture Notes on Computer Science, June 2006, pp. 194–211.

[26] Simatic M., Astic I., Aunis C., Gentes A., Guyot-Mbodji A., Jutant C., and Zaza E., "'Plug: Secrets of the Museum': A pervasive game taking place in a museum," in Entertainment Computing - ICEC 2009, Eighth International Conference, Paris, France, September 3-5, 2009, Proceedings, ser. Lecture Notes in Computer Science. Springer, September 2009, pp. 302–303.

[27] "PLUG: Play Ubiquitous Games and play more," http://cedric.cnam.fr/PLUG/ (last access in July 2012), August 2009.

Chapter 3

RFID Applications and Challenges

Ming-Shen Jian and Jain-Shing Wu

Additional information is available at the end of the chapter

1. Introduction

Radio Frequency identification (RFID) is the popular wireless induction system [1-7]. The same as the general bar code identification, each RFID tag in an RFID system is assumed that equips a unique ID (UID) itself. A standard RFID system is consisted of Tag, Reader, and Application. When an independent RFID tag approaches the RFID antenna, the induction between RFID tag and antenna happens. The RFID antenna reads the information and content recorded in the tag. Then the information is translated into the computational data by the RFID reader. Due to the portable RFID tag and untouched data transmission, many local or small area wireless applications for track and trace based on RFID systems were proposed.

RFID today is the popular wireless induction system [2-3,8-11]. Each RFID tag in RFID system is given a unique ID (UID) which records the on demand information. When an independent RFID tag approaches the RFID antenna, the induction between RFID tag and antenna happens. The information and content recorded in the tag is transmitted to the RFID antenna and translated into the computational data. Following up the data translation, the tag recognition can be completed and related applications are provided.

The RFID applications about agriculture that are now in widespread use such as the Animal Identification, the Product Record, and the Manufacturing process management. First, users identify products or materials via the tag and the reader of RFID system, and followed that recorded the data of products or materials on database in foregoing proposals. In a subsequent process, an especial function chose some suitable data from databases for analyzing, integration and description. This process helps users to understand the position and situation about products or materials. Therefore, by using the RFID, the main contributions in agriculture can be described as follows:

1.   The data of manufacturing record was transformed from artificial process to electrical process.

2.   Increased management (such as Inventory or Supply Chain) and production planning and scheduling efficiency than before.

3.   Reduced the cost of operation and increased the economic effect.

4.   Supply product safety information to customers to refer.

Many tracking applications based on ubiquitous computing and communication technologies have been presented in recent years such as RFID systems [4,5,7]. Therefore, RFID can be used to trace objects and asset worldwide. In addition, some warehouse systems or supply chain management systems can be combined with RFID to form goods tracking systems. The tracking systems help enterprises to manage their raw materials and products that reduce the cost of operation budget. However, more and more applications of RFID system that were introduced by people, and that the agriculture is the one of them.

Due to the popularity of RFID, many local or small area wireless applications were also proposed. The RFID tags were proposed to be used in hospital or health care [12-15]. Patients should always wear the RFID tag is designed for identification. The patient's current location and condition is monitored every time and everywhere within the hospital. It means that patients are under cared even an emergency state happens. Some entrance guard systems are also based on RFID system. The RFID ticket or RFID card [2, 3, 8, 10] is used to identify that a user is legal or not. According to the short-distance wireless signal, the RFID tag users can be monitored within the specific area. In other words, the RFID systems are generally used to be the hardware identification in many applications. In opposition to using the RFID system as the hardware identification, many software applications adopt software encryption as the identifications to protect the intellectual property of the applications or files. Considering the serious situations of pirate, intellectual property protection is important and becomes a famous issue.

Password protection is the popular encryption method to protect the applications. Each application or file of software is assigned an on demand given serial numbers or calculation function. People who use this application have to input the correct serial number then enable the application. Considering today's applications, personal multimedia services or software applications are popular. Customers use the personal multimedia devices such as MP3, PDA, iPod, Laptop, etc., to download the multimedia or application files from the server or website via Internet. In other words, many files or data are disseminated and exchanged via Internet. In addition, many hackers can crash the software encryption with fewer costs (Only program tools or applications needed). It makes that the piratical files are transmitted widely and the protection of intellectual property exists in name only.

For the purpose that the right of intellectual property and the right of the valid users are further protected and maintained, integration of the software and hardware encryption is needed. Since each RFID tag with a unique ID (UID) which records the on demand information can be used as the individual identification, the small and cheap RFID tag can be considered as the hardware/software encryption/decryption key corresponding to the files or applications. In the next section, we give some descriptions for related RFID application and system.

2. Related application and system

Some researchers presented that the embedding RFID can be plugged into a small device such as handheld host [1]. The handheld device users can plug in the SD or CF interface of reader card. Hence, the users can scan and induct the RFID tag everywhere. In other words, to integrate the RFID system hardware into the mobile devices is practicable. Furthermore, the RFID system including RFID induction antenna, RFID parser and reader, RFID tag, etc., today is cheap. In addition, the RFID hardware including antenna and reader is not only cheap but also can be a PnP device. It means that the RFID hardware can be used as a normal user device such as the card-reader.

2.1. RFID encryption and decryption for intellectual property protection

2.1.1. Application

Since the RFID systems are popular and ripe for distinguishing treatment of individual target [16, 17], the unique characteristic or identification of RFID can be the solution of intellectual property protection. Many researches proposed the possible way to protect the intellectual property, products, or applications. In some applications [18], the RFID chips are embedded in the cap of bottle. The medicine can be differentiated between fake and true. In addition, the RFID chip can be placed in the CD or DVD disk. The CD-ROM can accesses and reads the information of the RFID for valid identification check. Only the CD or DVD with the authorized RFID can be played. Although the content is protected, the self-made content that burned in the CD-R/RW or DVD-R/RW may not provide the authorized RFID information. In other words, the private, non-business, or free digital content made by the individual may be limited and cannot be transmitted free. In addition, even the CD or DVD disks are protected, the digital content such as files or data still can be copied from the disk to other devices such as hard disc or MP3 player. Therefore, how to separate the right of the digital content for each user and how to protect the digital content from illegal use become the important issues.

Due to the demand of existed system integration, some applications related to *RFID Encryption and Decryption for Intellectual Property Protection* includes: PnP Middleware, RFID Hardware, End User RFID Device and End User RFID Tag, and Encryption/Decryption Procedure. The system framework is shown as follows.

For a normal user, there are two types of RFID devices for the encryption/decryption on RFID system (E/DonRFID system): End User RFID Device for digital content or multimedia information gaining, and End User RFID Tag for indentifying the legal user.

E/DonRFID not only provides the RFID based protection procedure but also includes the Encryption/Decryption method based on RFID character. The encryption and decryption can be implemented by hardware or software solution. The original digital data is encrypted by 1) hardware, 2) software, or 3) combination of hardware and software. Corresponding to the encryption method, suitable RFID tag of user for decrypting is needed.

Since three possible ways to protect the digital content are proposed, for the end users, there will be at least three possible states and method of *E/DonRFID Encryption/Decryption*, to gain the protected digital data, shown as follows:

1. Encryption and Decryption by Hardware and Software combination,

2. Encryption only by Hardware with Hardware and Software combination Decryption

3. Encryption only by Software with Hardware and Software combination Decryption

4. Encryption only by Hardware with Hardware Decryption

5. Encryption only by Software with Hardware Decryption

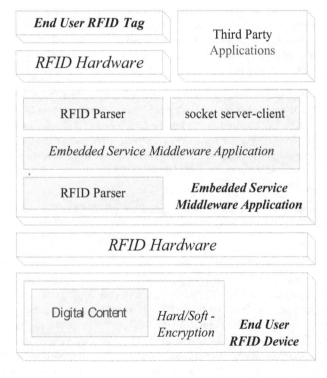

Figure 1. The framework of RFID Encryption and Decryption for Intellectual Property Protection

2.1.2. Method

First, depending on the digital content storage hardware such as CD-ROM disk, the commercial RFID tag can be embedded into the disk when the disk is made. According to the characteristic of RFID tag, each RFID tag can be set with different individualities. The different encryption code, unique ID, information of the digital content, or authentication serial number can be recorded in the RFID tag. In addition, the RFID tag embedded in the disk is not rewritable. Hence, different disks equip the different IDs of RFID tag. When the RFID reader inducts the tag, the information about this storage can be scanned and presented. In other words, only the digital storage with the valid RFID tag is legal and true.

Second, since the content or data are digital, these software, content or data, can be encrypted as the secret codes or cipher. The key for encryption and decryption can be recorded in the RFID tag. Without the specific key, these secret codes or ciphers cannot be recovered as the original data. In other words, the digital content that recorded in the storage device (such as CD-ROM disk) can be secured. The decryption key can be recorded in the RFID tag embedded in the storage or a palm RFID tag (such as a RFID toy).

For the end users, *End User RFID Device/Tag* is used. The storage, whether hardware (CD-ROM) which includes the encrypted digital content, or software (files or ciphers), is called *End User RFID Device*. If the *End User RFID Device* is hardware, the third party *RFID Hardware* can induct the RFID tag embedded in the hardware. After identifying the *End User RFID Device*, the application or user can execute and read the digital content if only *Hardware - Encryption/Decryption* is used.

According to three possible states, the end user must have the decryption key for executing the digital content. In this paper, the hardware (RFID tag) or software for the decryption key is called *End User RFID Tag*. After identifying the *End User RFID Device*, the end user has to provide the *End User RFID Tag* for the *Embedded Service Middleware Application*. Only the information or password of *End User RFID Tag* is correct and can be used to gain the secured decryption key which recorded in the *End User RFID Device*, the digital content recorded in the *End User RFID Device* can be presented.

Considering that the three possible states are based on the RFID induction, the *RFID Hardware* is divided into two types of equipments: for *End User RFID Device* and for *End User RFID Tag*.

According to the three possible ways to protect the digital content, when the protection is based on the combination of *Hard/Soft- Encryption/Decryption* and Only *Hardware-Encryption* with *Hard/Soft –Decryption*, the *RFID Hardware* for *End User RFID Device* is needed. Due to that the digital content is protected by the RFID tag embedded in the hardware, the information recorded in the tag has to be inducted before using. For example, if a tag is embedded in the CD-ROM disk, the user should have a CD-ROM driver with the *RFID Hardware* when reading the disk. In other words, if the protection is based on the hardware belongs to *End User RFID Device*, the corresponding reader with *RFID Hardware* is necessary. The *RFID Hardware* can be embedded in the CD-ROM driver, reader, or other multimedia devices.

In opposition to *End User RFID Device*, when the decryption is based on the *End User RFID Tag* key, end user has to own the valid RFID tag for decrypting the digital content. For example, the decryption code is recorded in the RFID tag of *End User RFID Device*. However, the decryption code is secured by the password which locks the data slot of RFID tag. Without the correct password, end user cannot gain the decryption code that secured in the RFID tag of *End User RFID Device*. To provide the password, the end users should have the *RFID Hardware* such as the USB-RFID reader, etc.

To manage the RFID information, *Embedded Service Middleware Application* is proposed to parse the information from the *RFID Hardware*. Due to that there are different RFID product, an RFID parser is needed for analyzing and parsing the information from *RFID Hardware*. After gaining the requirements or response, the *Embedded Service Middleware Application* searches the corresponding applications and passes the information. Figure 2 presents the framework of *Embedded Service Middleware Application*.

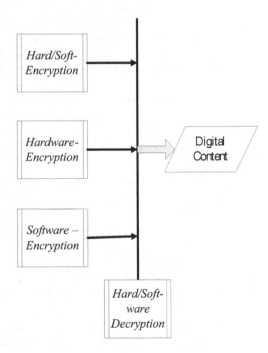

Figure 2. Possible ways to gain the protected digital data

The *Embedded Service Middleware Application* implements the socket server-client structure for communication with other existed or third party applications. The information comes from the *End User RFID Device*, such as specific password-requirement, will be recorded in the database of middleware application. The requirement will be maintained based on the on demand limitation of the period of validity or when the *End User RFID Device* is removed. In addition, when an end user tries to gain the digital data from the *End User RFID Device*, the middleware application request the end user for the password. After receiving the password, the middle application transmits the password and tries to gain the decryption code. If the password is correct, the decryption code will be transmitted to the user application such as multimedia player, etc. Otherwise, the digital content cannot be decrypted and used. Therefore, only the two conditions: 1) the key information of *End User RFID Tag* matches the password requirement of *End User RFID Device*, and 2) the decryption code is correct in decrypting the digital content are satisfied, the user can gain the information from the *End User RFID Device*

2.2. Location aware public/personal information services based on embedded RFID platform

2.2.1. Application

Many researches proposed before presented the importance of providing information and services related the user's location to each person. Some researches assume that there are GPS devices or module included in the users' mobile devices. Then, according to the information of GPS (GIS) [19-22], the location aware or related information or services are provided to the mobile user trough the wireless network. Although the GPS provides the accurate location of users, most users indeed needs the approximate local-area-aware information. The accuracy of location such as longitude and latitude is not the main issue. Furthermore, not everyone can equip the GPS.

Hence, in addition to GPS, according to the orientation made by the station of wireless cellular system [23], the related information according to the user's location can be given to the user via cellular system. Each cellular phone user can be served directly by the telecommunications companies. If a user is served by the specific wireless base station, the information related to the coverage area of this base station are given to the phone user.

Since the RFID system is popular and generally implemented, many researches [24-30] tried to integrate the RFID to and applied RFID technology to context-aware systems. However, in [31], what kind of the context, the corresponding context services, and the context-aware RFID system are important to be provided for user is still an issue of the existing system. In addition, to integrate the existed system such as information service and payment system become the important topic.

Not only supply the public services but also give the personal services, the context aware researches [32-34] were also proposed. Research in [35] was proposed that considering the user's related location. Hospital or health care RFID systems [12-14,34] for monitor the tag users were also proposed. A designed RFID tag is given to each user such as a patient. Each patient should always wear the RFID tag every time and everywhere. Hence, the patients' current information such as location and health conditions are monitored by the hospital. In addition, some entrance guard systems are also based on RFID system. The RFID ticket or RFID card [2, 3, 10, 36] is used to identify that a user is legal or not.

The services and information of user-location-related public places such as the museum [37] are provided. According to the requirement of users, different services are given through wireless network or cellular system to different users even they are in the same places.

Hence, a realistic application such as Location Aware Public/Personal Information Services based on RFID is proposed. By using the location-aware RFID application, the main contributions are :

1. Users of location-aware RFID application can communicate and gain the information corresponding to the users' location through the RFID tag. The handheld devices with RFID reader can also manually obtain the extra or required or local information and services.

2. The efficiency of system management and service utilization can be improved, information can be the digital multimedia and updated immediately,

3. The location-aware RFID application can be embedded in other similar service systems and hardware. The proposed service system can be included in the existed information center or server. The additional cost for integration can be reduced.

4. The service object and function can be various. For example, the location-aware RFID application provides not only the public or general information services to every system user, but also the deferential personal service to individual location-aware RFID application user.

5. The ticket and payment services can be integrated into location-aware RFID application service system. Users needn't to bring too many identification devices or cards. All method of payment and public or personal services are integrated into one RFID tag and location-aware RFID application service system.

2.2.2. Method

The system structure is shown as Figure 2. The *Embedded Service Middleware Platform* is the main system to manage the internal and external system connections. The RFID API and parser are included and provided to communicate with the third party RFID system. The *Embedded Service Middleware Platform* also makes the information connection to other business management system or database via software API. In addition, the related information to the RFID tag inducted is presented by user interface.

For the end users, *End User RFID Handheld Facilities* consists of two appliances: *end user RFID tag* and *end user device with RFID System*. A user can use a given readable and re-writable RFID tag or a handheld device such as PDA which equipped a RFID system to gain the required public/personal services. In Figure 2, the user handheld device also equips the RFID system, RFID API, and parser to scan and induct the commercial RFID tag. The communication and the data transmission between the handheld device and server can be established via 1) Internet, 2) server-client socket, 3) a user RFID tag, or 4) a readable and re-writable RFID tag. A user can view the information or obtain the services via user interface (UI) presented by server or the user handheld device.

The other business management systems in the framework can be the third party developments and independent of the whole system. When the user approaches the RFID system at the
specified area, the induction and communication between end user RFID tag and antenna of
RFID System is automatically established. A RFID reader will parse the signal into the digital
and computing content. Then, the *RFID System* transmits the information obtained from the tag
to the *Embedded Service Middleware Platform* via Internet. According to the RFID information,
the *Embedded Service Middleware Platform* searches for and provides the specific personal service recorded in local area server according to the on demand conditions of the user. Moreover,
the information or services can be updated or provided from the main database via Internet
connection. Then, the user can obtain the public/personal information from the user interface.

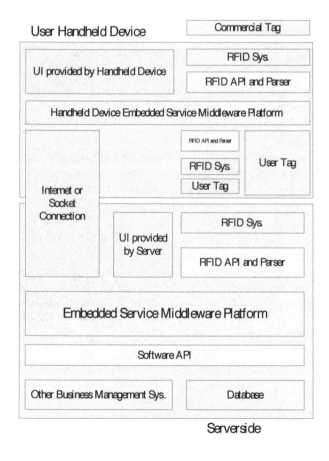

Figure 3. The whole framework of location-aware RFID application service system

In the location-aware RFID application service system, the RFID antennas and reader are deployed 1) at the specific area or location such as the entrance of the rapid transit system or the information service machine, or 2) within the handheld devices such as PDA or mobile phone. When a user is given a readable RFID tag, the related information or the user's on demand service conditions about the user is given by himself and on demand recorded in the database. When the user requires the local area public or personal services, the user should be at the tiny induction area such as a local area information center or a service station. Then, the RFID system placed in the specific area inducts the RFID tag and gain the information such as UID from the RFID tag. The reader of RFID system then sends the information to the local area server via Internet.

After receiving the information and parsing the message from RFID system, the content of RFID tag can be identified. If *end user RFID tag* is used, the *embedded server service middleware* can search and present the local information such as local area shopping information, traffic information, or the customization information, recorded in local database that match the on demand conditions of the RFID tag user. In other words, the RFID user can be directly served with sufficient local area related information. If other further information needed, the *embedded server service middleware* can send the user's request to the remote main server to obtain the requested service or to the other business applications via Internet for extra service obtaining. At last, the user can gain the location-aware information or services via user interface. In opposition to *end user RFID tag*, when a user of *end user device with RFID System* actively scans the RFID tag of the commercial advertisement, the handheld device can send the scanned RFID tag information via wireless network or cellular mobile system to the local area server with *embedded server service middleware* embedded. Then, as the procedure of *end user RFID tag*, the *embedded server service middleware* searches for the requested services and transmits these services to the user's handheld device by wireless network or cellular mobile system.

In addition, the users can use *handheld device middleware application* to select the tag content recorded in handheld device database if needed. Then, the RFID API controls the RFID system embedded in the handheld device to re-write the content (such as UID) of the tag of the handheld device. At last, the RFID content requirement from other business applications or systems can be provided through the RFID tag of the handheld device.

In addition, the database can record the history of the user's requirements. The statistic user requirements can be used to classify that what kind of the service the user requests most. Next time the system can provide the personal services according to the classified results. In other words, the users can be served with the services they most pay attention to.

The real test and verification is implemented as Figure 3. The implementation shows one shoot of the verification. When the user approaches the on demand placed RFID system, the *Embedded Service Middleware Platform* automatically presents the information corresponding the content or user's related information recorded in the RFID tag. For example, if a Taiwanese uses the RFID tag, the presentation of local area server will be based on traditional Chinese. But the local area server will provide English when a native English speaker user his own RFID tag respectively.

Figure 4. The environment of test and verification

2.3. RFID applications on supply chain management

2.3.1. Application

Existing RFID applications on supply chain management (SCM) can record something about materials, goods, and products during production [38, 39]. An integrate system with RFID and SCM also can supplies new value-added services such as products secure protection and to query products record [40]. And integrating promising information technologies such as RFID technology, mobile devices-PDA and web portals can help improve the effectiveness and convenience of information flow in construction supply chain control systems [41]. In addition, RFID can be use in a lifecycle of a product to reduce the time which spend to find a product. Therefore, RFID is a technology to reduce the time to identified objects that can improve automation in the traceability management of Supply Chain.

In Warehouse management, many companies have used RFID technology replaced the Bar Code or QR Code as recognition of the key features. Because of the Bar Code and QR Code are limited in existing format companies decided to select a solution improving automation in warehouse management. Then the RFID technology is an efficient technique to solve that problem. Company integrated RFID technology with warehouse management that is not only an electronic process solution but can provided customers with new services such as location information of products, search stock of products, and provide inventory information [42, 43].

In healthcare, RFID also can use to trace patients, blood sampling, drop management, etc. Kumiko Ohashi, SakikoOta, LucilaOhno-Machado, and HiroshiTanaka [44] developed a smart medical environment with RFID technology. This research used two types of frequency on tracking system for tracking clinical intervention such as drug administrations and blood tests at the patient bedside. Furthermore, Chung-Chih Lin, Ping-Yeh Lin, Po-Kuan Lu, and Guan-Yu Hsieh [45] proposed a healthcare integration system for disease assessment and safety monitoring of dementia patients. The proposed healthcare integration system provides the development of an indoor and outdoor active safety monitoring mechanism.

Hence, due to that the RFID technology could provide some services with auto-identify such as administration of drops and samplings, safe monitoring of patients, process control in medical. These new type services can reduce the search time in administration of drops and samplings and human error in medical process. The major advantage of using RFID technology in medical is reduced the human error.

However, RFID applications on Supply Chain Management and Warehouse Management were provided static information as previously noted which helped to deal with problems after accidents. Information on RFID systems was lacking warning data of preventing accidents [5]. To summarize, both Supply Chain Management and Warehouse Management are increase economic values of product. Those operations of management can be help to support product safety information and attribution of responsibility information for customer and enterprise. When enterprise want reduce possible impaired factors to improve the value of product, the first thing should to do is disease management. In addition, disease management has two important methods which to find out the initial pathogen and reduce the spread of pathogens and the infection rate. The spread of pathogens and the infection rate decide the effect areas and damages. Because of enterprise can economic damage control by detect symptoms of product at early stages that will help to reduce cost of operation by itself.

For example, to improve the efficiency of management in cultivation, an *RFID Based Fuzzy Inference Algorithm for Disease Warning and Tracking via Cloud Platform* is proposed. Users could manage the cultivation history, related bio-information, and possible disease tracking. The proposed system modifies the traditional cultivation management system by several fields: 1) first, to shorten the processing time of object recognition in production operation; 2) second, to establish electronic records of production in production management systems; 3) third, to integrate supply chain managements with a central server and provide real-time environment monitoring and plant disease management services for users; and 4) last, to establish an information platform to share with users.

Due to that the contents storage in the memory of RFID tag can be changed when users need. Furthermore, RFID can apply in recognition and also can work in hostile environment such as wet and dirty [26]. RFID provides large read range (or induction range) than Barcode and QR code. Therefore, the RFID system can help to efficiently identify object which equips RFID tag even in non-uniform position. Besides, RFID tag is rewritable. User can remove or rewrite the content of RFID tag when the induction happened between the RFID tag and the RFID reader. In order to overcome the environment factors of cultivation and the objects size, RFID is the solution that can suitable to solve these problems.

In recent years, transportation becomes faster with long distance and also causes more areas infected by disease more easily. After infectious disease influencing a mounts of areas, disease management is more complex and ineffectual [6]. Therefore, quick disease control and prediction is important since it could help to reduce the cost and complex of disease management. Hence, effective disease data tracking and collection of pathogens is necessary.

Due to portable RFID tags and non-touched transmission, local area wireless application about disease management for tracking and collection data based on RFID system is proposed. In addition, an information platform which collects data from everywhere and stores the data in its database for the members of supply chain is needed and important. Every platform user can query and access some information from an information sharing platform via network. An information platform can store a lot professional data of a particular field. The platform also can integrate information from each region and has more powerful computing for more information services.

2.3.2. Method

The proposed application system structure is shown as Figure 4. The system infrastructure includes RFID system such as RFID tag, RFID reader, mobile RFID device, etc., and software framework such as database, *Environmental Affection Evaluation Method*, and *Disease Tracking Service* in cloud. The user application layer indicates the mobile RFID device which is used for inducting the RFID tag of local objects such as crops or livestock. The information read from the RFID tag by the RFID reader will be transmitted to the corresponding application and database in cloud. Considering the real implementation, *Environmental Affection Evaluation Method* and *Disease Tracking Service* can be established as the middleware of the whole system or the corresponding applications in cloud. When a user wants to query the information or obtain the disease warning, the proposed *Environmental Affection Evaluation Method* and *Disease Tracking Service* can notify the client user via network from cloud platform.

Generally, an RFID system includes RFID Tags, RFID Readers and Application programs. An RFID tag is a digital storage device that used for identification and information recording. A Reader can access, read, or write data into RFID tags through electromagnetic induction. A user can only use the RFID tag without power consumption. In addition, the mobile device used in RFID system can be a Person Digital Assistant (PDA), a Person Computer (PC) or a laptop (Notebook), which executes the reading and writing actions via RFID systems (include software API and hardware). The middleware mainly manage and deal with the RFID event such as the RFID information sent to or from other systems. After receiving the message from RFID readers, the content of RFID tag can be identified. Then the RFID information will be transmitted and recorded in the database in cloud by *RFID Event Processing*. In addition, the corresponding information sent by *RFID Event Processing* is also presented by the user interface. The *RFID Event Processing* also properly manages and provides the information service for *Environmental Affection Evaluation Method* which analyzes and evaluates the affection degree of the environmental factors.

After inducting the RFID, the information can be transmitted to the corresponding applications and recorded into database in cloud. Each RFID tag will establish an individual object history about the location, resident time, environment state of the RFID tag when it was storaged, etc. To trace and track the potential diseased objects, the *Disease Tracking Service* for client users is needed.

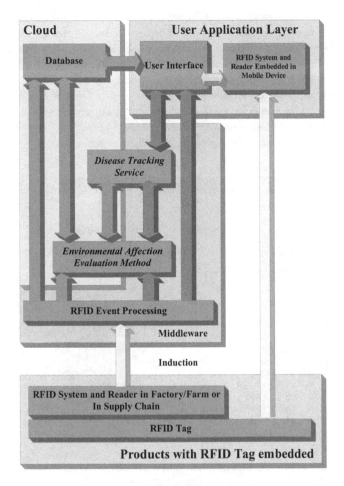

Figure 5. The structure of RFID system with Cloud environment

Via *Disease Tracking Service*, customers or client users can obtain active and passive warning message. First, when a customer or client user uses the mobile RFID system to query the information, the corresponding record in the past will be checked and sent back from the database in cloud platform. Then, the members of the supply chain can monitor and manage the state of objects. Second, if the specific object (crop or livestock) is found to be diseased, this object will be marked as the dangerous object. Then, according to the object history, the route, location, etc., where this dangerous object ever passed will be traced back. According to the structure presented in Figure 2, since the object history is established in cloud, the members of the supply chain can exchange the information via cloud. Therefore, the history of the dangerous object can indicate the information about place, location, resident time, etc.

Therefore, if the *Disease Tracking Service* finds the dangerous object, the corresponding history, members of supply chain, and the potential diseased that evaluated by *Environmental Affection Evaluation Method* can be notified and traced. In other words, no matter where the objects are, the *Environmental Affection Evaluation Method* can always give the probability value of objects which indicate the potential diseased probability. By using the *Disease Tracking Service*, the location, warehouse, manager, etc., will be notified that how many objects with the different and individual potential diseased probability currently reside at or ever passed the place. Therefore, the object with high potential diseased probability can be discovered in early phase. Figure 5. presents the implementation of the system.

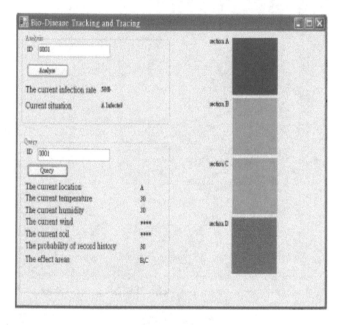

Figure 6. The implementation of tracing and tracking system based on RFID

The tracing and tracking system based on RFID system not only provide original services in supply chain or warehouse which includes the records of products for managing and querying, but also provide new services such as to obtain the potential probability of diseases and send the warning sign for users. Furthermore, the system can derive the possible disease infecting area for users to control and update the latest information of production for users to track and trace. In addition, the verification shows that the proposed system is realistic and can provide the public and personal services automatically. By using this innovation RFID system, users could get most information to prevent disease in agricultures that helps users to reduce the cost of production, control the range of disease occurrence, and providing a warning for disease prediction.

3. Discussion

Radio Frequency identification (RFID) is the popular wireless induction system [7] [2, 3, 8, 30]. Each RFID tag in an RFID system is equips a unique ID (UID) itself. UID can help to shorten the identification time for individual object recognition. In general, there are several methods to achieve aim of Automatic Identification such as Barcode and QR code. However, due to that the Barcode and QR code have the limitation in environmental affection such as wet or water, to maintain the usability and the reliability of the Barcode or QR code is too difficult. Therefore, using RFID can be the solution which provides the distance induction with better characteristics such as anti-water and rewritable memory.

A standard RFID system consists of Tag, Reader, Middleware, and Application. When an independent RFID tag approaches the RFID antenna, the induction between RFID tag and antenna happens [9]. The RFID antenna reads or obtains the information and content recorded in the tag. Then the information is translated into the computational data by the RFID reader. Due to the portable RFID tag and untouched data transmission, many local or small area wireless applications for track and trace based on RFID systems were proposed [2, 7, 8].

RFID reader can access data of RFID tag and transmit the content from RFID tag to middleware which is a necessary component in RFID application system. The middleware is also the interface software that connects new RFID hardware with legacy enterprise IT systems [36]. Middleware is used to route data between the RFID networks and the IT systems within an organization. It merges new RFID systems with legacy IT systems.

RFID Reader is also called Interrogator. The RFID reader can read and write data of RFID tag via radio frequency. RFID readers can classify serial reader and network reader according to connection interface.

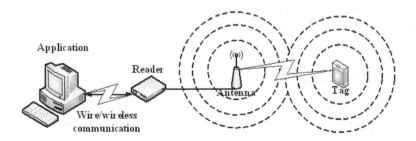

Figure 7. Operation concept of RFID system

In Radio Frequency Identification, there are four standard frequency ranges used: Low Frequency (LF), High Frequency (HF), Ultra-HF (UHF), and Microwave respectively. Frequency decides the reading distance of RFID devices and the interfered with environment. Figure 7 shows the shows the conditions and factors related to the frequency. Higher frequency RFID tag has longer induction distance, higher data rate, and smaller Tag size. On the contrary, lower frequency RFID tag has shorter induction distance, lower data rate, and bigger tag size. In addition, higher frequency RFID tag has bad performance when tag near metal or liquids. In this thesis, the proposed system selects low frequency RFID tag because that the cultivation environment is wet and dirty. Low frequency RFID tag has better performance than high frequency RFID Tag at cultivation environment.

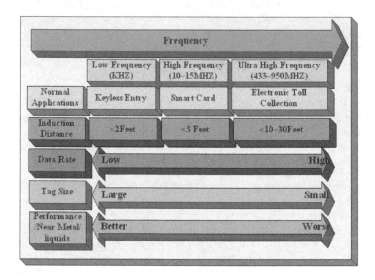

Figure 8. RFID frequency specification chart

There are two main types of RFID tag, active and passive. Active tag is a powered tag which equips battery and actively sends signal itself. In opposition to active tag, passive tag without battery can only send signal when the tag is inducted. Table 1 presents an introduction for RFID Tag.

Freq. Effect	Low Frequency(LF)	High Frequency(HF)	Ultra-HF (UHF)	Microwave
Usable Frequency	100~500 KHz	10~15 MHz	433~955 MHz	1GHz~
Common Frequency	125 KHz 135 KHz	13.56 MHz	433 MHz 868~950 MHz	2.45 GHz 5.8 GHz
Power Type	Passive	Active / Passive	Active / Passive	Active / Passive
Reading Distance	Short range	Short range	Longer range	Longer range

Source: EPCglobal.(http://www.gs1tw.org/twct/web/EPC/index.jsp)

Table 1. The relational table of RFID frequency and instruction.

The RFID technology is affected by several factors such as frequency, energy, and environment. When users decide to use RFID in a particular place, users have to select the fittest specification of RFID at first. Because the fittest specification of RFID can make higher performance in the efficient of RFID application system, to select the fittest specification of RFID is important. Furthermore, the cost of RFID is an important factor for users to select.

4. Challenges

4.1. Cost

Today, based on the behavior of customers, the history or record of the production including the delivering is not the most important thing. Most consumers care about the price, expiration date, or packing of the goods. In other words, although the RFID can enhance the management of logistic, to embed the RFID technology into current system also cost a lot. For example, one RFID tag for paste once (non-reusable) may cost $0.8. However, a bottle of water may also cost about $1. Most consumers would not to pay almost twice payment to obtain the information which they do not care about. Comparing with barcode or 417-barcode, the RFID tag costs more than barcode or 417-barcode which can be printed by a printer. In other words, to identify an object via RFID tag costs more than using barcode.

In opposition to once-using RFID tag, to reuse the RFID tag may be a solution. A plastic card where RFID embedded can be used for identification or payment. For example, in Taiwan, most mass transit systems can be paid according to the RFID card that pre-registered and sold to the consumers. Each consumer can use the RFID card to pay the consumption in convenient store, fee of the parking lot, and use the card as a ticket for railway or MRT system. Due to the convenience, most consumers have at least one RFID card for payment. However, the security for financial application is very important. Although there are some security method proposed, no algorithm or system can provide the 100% promise that the security method is always safe.

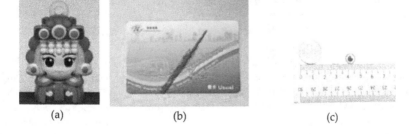

(a)   (b)   (c)

Figure 9. The sample of RFID Tag; a.3D Toy, b. plastic card, c. button tag.

In addition to the RFID card, various styles of RFID tag may be required by different consumer. According to the environment, the RFID tag may be designed to anti-water or anti-wet. Some RFID tags need to be applied to the metal environment such as containers. Due to the characteristics of RFID, suitable frequency of RFID corresponding to the environment should be selected carefully. In other words, the cost of different RFID may be also different according to the applications and environment. Furthermore, different shapes of RFID tag also costs. For example, the cost of an RFID tag embedded into a 3D toy is much more expensive than that embedded in a plastic card.

4.2. Hardware and integration

The RFID Antenna is the main component for RFID tag induction. The antenna continuously spreads the electromagnetic wave. The energy is transmitted to the RFID tag. After induction, the RFID Antenna also receives the signal from the RFID tag.

After receiving the signal, the RFID Reader translates the signal into the digital data such as the UID of this RFID tag. Then, the RFID Reader sends the digital data to the corresponding systems or applications. To implement the RFID system, not only the RFID tag but also RFID reader and antenna hardware should be considered. Due to the design and product limitation, the RFID antenna cannot dynamically change. Therefore, similar to the RFID tag, the environment of the RFID system affects the size and cost of the antenna. The size of the antenna increases, the costs also increases. In addition, considering the implementation environment, to place the antenna at the suitable location for signal receiving also affects the performance and costs. Therefore, to integrate the RFID system with the existed system, some additional problems may need to be overcome.

The existed applications or systems should include or integrate the RFID system. In other words, to integrate the RFID system, the original working practices of the existed system may be changed which needs extra costs. For example, to identify goods in warehouse, original applications or systems may only need the manual operation. However, to integrate the RFID system, some infrastructure such as the placement of RFID antenna, wire line for connection between antenna and reader, and the establishment of RFID reader and system server are required. In other words, the extra costs of RFID infrastructure are needed. In addition, some existed systems are based on mechanical operation without too much intelligent analysis. For example, a car parking lot only needs to open the gate when a car approaching or according to the teleswitch. When integrated with the RFID system, the RFID antenna should be placed in front of the gate for induction. All the cars to the parking lot should present the RFID tag given on demand. In addition, the RFID reader should be used to analyze the signal information from the RFID tag. The server which includes the database should be used to judge whether the gate should open or not. Although the RFID system enhances the automation with less manual operation, some extra costs and delay may also happen. Therefore, the benefits of RFID system integration such as automation, information exchanging with third party applications, etc., are very important. Only when the benefits or additional new functions overcome the extra costs of RFID system integration then the integration of system will be used.

4.3. Plug and play middleware

In the RFID systems or applications, there are two partitions: RFID devices (includes RFID tag, antenna, and reader) and other devices or systems. Therefore, the application or middleware for communicating these two parts is needed. When using the RFID device, the third party systems or applications should obtain the information from RFID devices.

Due to that there are many types of RFID hardware, the application program interface (API) for the communication between RFID Hardware and different third party applications is needed. In addition, the end user's devices are also various. Hence, the plug and play middleware for different hardware and applications is important.

To manage the RFID information from different RFID Hardware, and the communication with different applications, the Plug and Play Middleware is proposed. To realize the concept of Plug and Play, the proposed middleware has to manage the information from the all possible third party RFID Hardware, deal with and parse the information, and then provide the required information to the corresponding applications. Therefore, the main purposes of the proposed Plug and Play Middleware are:

1. to parse the information from the RFID Hardware. Due to that there are different RFID product, the RFID parser is needed for analyzing and parsing the information from RFID Hardware. The information about UID, password, etc. will be parsed as the string for the further execution of applications. In this paper, two possible parsers are established. First, the Plug and Play Middleware provides the remote procedure call (RPC) function for the third party RFID Hardware. The UID of the RFID tag inducted by the RFID Hardware will be formulated as the string. In addition, the password or requirements for further information such as decryption code recorded in the End User RFID Device can be provided by the remote procedure call function. Second, for general communication, the Plug and Play Middleware also provides the sever-client socket link between the RFID Hardware and the middleware. In other words, even the RFID Hardware cannot implement the remote procedure call, depends on sever-client socket link, the information can be transmitted between Plug and Play Middleware and RFID Hardware.

2. to provide the application program interface (API). Since the RFID Hardware may not directly communicate with the applications, the Plug and Play Middleware has to implement the corresponding API for other third party applications or software.

Furthermore, the Plug and Play Middleware also should implement two possible APIs: the external procedure call and network communication. If the application is embedded in the Plug and Play Middleware, the external procedure call sends the required information to the specific application. In addition, some communications of the related applications such as database query are also established by the external procedure call. Then, the Plug and Play Middleware deals with the results from the external procedure call. In opposition to external procedure call, for the concept of Plug and Play, normal network communication should also be implemented. Most third party software or applications can communicate with the Plug and Play Middleware via sending the information in string format. For example, if the third party application requires the further checking, the Plug and Play Middleware sends the required information such as UID to the server via Internet. To reduce the cost for communicating with different third party applications, the unify data storage format is necessary. Therefore, the eXtensible Markup Language (XML) can used as a data exchange standard. After obtaining the response from the server, the Plug and Play Middleware can acknowledge the third party application. At last, the corresponding services can be presented.

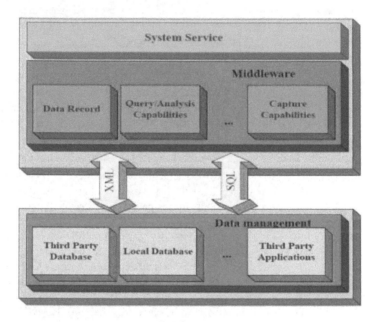

Figure 10. The structure chart of data exchange in Database management

5. Conclusion

In this chapter, we show applications and systems based on RFID technology which integrated into the existed service systems. The RFID technology can enhance the automatic management procedure. Identification and tiny information exchanging can be achieved. Individual or personal services can be provided to different consumers. However, to establish the RFID embedded systems and applications, the cost, convenience, feasibility should be considered. To adopt RFID system, some extra costs such as RFID tag and hardware should be overcome by the enhanced performance of management. In other words, to implement the RFID systems for the consumers, to enhance the convenience for consumers will be an important issue than the cost.

Author details

Ming-Shen Jian[1] and Jain-Shing Wu[2]

1 Deptartment of Computer Science and Information Engineering, National Formosa University, Taiwan, R. O. C.

2 Deptartment of Computer Science and Engineering, National Sun Yat-sen University, Taiwan, R. O. C.

References

[1] "Hoboken RFID-enables Its Parking Permits," RFID Journal, Jun 2006, http://www.rfidjournal.com/article/articleview/2421/1/1/

[2] "RFID Takes a Swing at Ticket Fraud", RFID Journal, Dec 2005, http://www.rfidjournal.com/article/articleview/2060/1/1/

[3] "Moscow Metro Tries RFID-Enabled Ticketing," RFID Journal, Feb 2007, http://www.rfidjournal.com/article/view/3049/

[4] A.S. Martinez-Sala, E. Egea-Lopez, F. Garcia-Sanchez, J. Garcia-Haro, "Tracking of Returnable Packaging and Transport Units with active RFID in the grocery supply chain," Computers in Industry, Vol. 60, Issue 3, pp.161-171, Apr 2009.

[5] T.C. Poon, K.L. Choy, Harry K.H. Chow, Henry C.W. Lau, Felix T.S. Chan, K.C. Ho, "A RFID case-based logistics resource management system for managing order-picking operations in warehouses," Expert Systems with Application, Vol. 36, Issue 4, pp. 8277-8301, 2009.

[6] C. Loyce, J.M. Meynard, C. Bouchard, B. Rolland, P. Lonnet, P. Bataillon, M.H. Bernicot, M. Bonnefoy, X. Charrier, B. Debote, T. Demarquet, B. Duperrier, I. Felix, D. Heddadj, O. Leblanc, M. Leleu, P. Mangin, M. Meausoone, G. Doussinault, "Interaction between cultivar and crop management effects on winter wheat diseases," lodging, and yield, Crop Protection, Vol. 27, Issue 7, pp. 1131-1142, 2008.

[7] Colin H. Burton, "Reconciling the new demands for food protection with environmental needs in the management of livestock wastes," OECD Workshop, Vol. 100, Issue 22, pp. 5399-5405, 2009.

[8] Beijing Olympic Games Prompts RFID Development in China, http://www.rfidglobal.org/news/2007_9/200709031653253861.html

[9] M.S. Jian, K. S. Yang, and C.L. Lee, "Context and Location Aware Public/Personal Information Service based on RFID System Integration," WSEAS Trans. on Systems, Vol. 7, pp. 774-784, Jun. 2008.

[10] Z. Pala and N. Inanc, "Smart Parking Applications Using RFID Technology," Proc. of 1st Annual RFID Eurasia, pp. 1-3, Sep 2007.

[11] M.F. Lu, S.Y. Chang, C.M. Ni, J.S. Deng, and C.Y. Chung, "Low Frequency Passive RFID Transponder with Non-revivable Privacy Protection Circuit," Proc. of WSEAS Inter. Conf. on Instrumentation, Measurement, Circuits, and Sys., pp. 166-169, April 2006.

[12] "Hospital Uses RFID for Surgical Patients," RFID Journal, Jul 2005, http://www.rfid-journal.com/article/articleview/1714/1/1/

[13] "RFID Hospital: Columbus Children's Hospital To Install RFID System From Mobile Aspects", RFID Solution Online, Mar 2007.

[14] "RFID trial tracks hospital equipment," http://www.computing.co.uk/computing/news/2168717/rfid-trial-tracks-hospital

[15] C.L. Lai, S.W. Chien, S.C. Chen, and K. Fang "Enhancing Medication Safety and Re-duce Adverse Drug Events on Inpatient Medication Administration using RFID," WSEAS Trans. on Communications, Vol. 7, pp. 1045-1054, Oct 2008.

[16] M.S. Jian and S.H. Hsu, "Location Aware Public/Personal Diversity of Information Services based on embedded RFID Platform," Proc. of ICACT'09, pp. 1145-1150, Feb 2009.

[17] Ming-Shen Jian, Kuen Shiuh Yang, and Chung-Lun Lee, "Modular RFID Parking Management System based on Existed Gate System Integration," WSEAS Trans. on Systems, Vol. 7, pp. 706-716, Jun 2008.

[18] http://www.ist.com.tw/

[19] C.S. Baptista, C.P. Nunes, A.G. de Sousa, E.R. da Silva, F.L. Leite, and A.C. de Paiva., "On Performance Evaluation of Web GIS Applications," Proc. of the IEEE 16th Inter. Workshop on Database and Expert Sys. App., pp. 497-501, 2005.

[20] M. Marmasse and C. Schmandt, "Locationaware information delivery with commo-tion," Proc. of 2nd Inter. Symposium in Handheld and Ubiq. Comp., 2000.

[21] D. Ashbrook and T. Starner, "Learning significant locations and predicting user movement with GPS," Proc. of Inter. Symposium on Wearable Comp., 2002.

[22] D.J. Patterson, L. Liao, D. Fox, and H. Kautz, "Inferring High-Level Behavior from Low-Level Sensors," Proc. of 5th Inter. Conf. on Ubiqutious Comp., 2003

[23] P. Bahl and V.N. Padmanabhan, "RADAR: An inbuilding RF-based user location and tracking system," Proc. of Infocom, pp.775-784, 2000.

[24] [24]K. Romer, T. Schoch, F. Mattern, and T. Dubendorfer, "Smart Identification Frameworks for Ubiquitous Computing Applications," Wireless Networks, pp. 689-700, 2004.

[25] S. Willis and S. Helal, "A Passive RFID Information Grid for Location and Proximity Sensing for the Blind User," TR04-009, 2004.

[26] G. Chen and D. Kotz, "A Survey of Context-Aware Mobile Computing Research," Dartmouth College Technical Report TR2000-381, 2000.

[27] J. E. Bardram, R. E. Kjar, and M. O. Pedersen, "Context-Aware User Authentication - Supporting Proximity-based Login in Pervasive Computing," Proc. of Ubicomp, pp. 107-123, 2003.

[28] C. Floerkemeier and M. Lampe, "Issues with RFID usage in ubiquitous computing applications," Proc. of Pervasive Comp., pp. 188-193, 2004.

[29] M. Crawford, et. al., "RFID Enabled Awareness of Participant's Context in eMeetings," Proc. of Pervasive Tech. Applied Real-World Experiences with RFID and Sensor Net., 2006.

[30] J. Bravo, R. Hervas, G. Chavira, and S. Nava, "Modeling Contexts by RFID-Sensor Fusion," Proc. of Pervasive Comp. and Comm. Workshops, pp. 30-34, 2006.

[31] J. Choi and E. Kim, "Using of the Context in RFID Systems" Comm. of the Korea Information Science Society, pp. 64-70, 2006.

[32] L. Han, S. Jyri, J. Ma, K. Yu, "Research on Context-Aware Mobile Computing," Proc. of 22nd Inter.Conf. on Adv. Infor. Net.g and App. –Workshops, pp. 24-31, 2008.

[33] L. Buriano, "Exploiting Social Context Information in Context-Aware Mobile Tourism Guides," Proc. of Mobile Guide 2006, 2006.

[34] M. A. Munoz, M. Rodriguez, J. F. Center, A. I. Martinez-Garcia and V. M. Gonzalez. "Context-Aware Mobile Communication in Hospitals," IEEE Computer, pp. 38-46, 2003.

[35] C. Ciavarella and F. Paterno, F., "The design of a handheld, location-aware guide for in-door environments," Springer Verlag Personal and Ubiquitous Computing, pp. 82-91, 2004.

[36] Z. Pala and N. Inanc, "Smart Parking Applications Using RFID Technology," Proc. of 1st Annual RFID Eurasia, pp. 1-3, Sep 2007.

[37] R. Tesoriero, J. A. Gallud, M. Lozano, and V. M. R. Penichet, "A Location-aware System using RFID and Mobile Devices for Art Museums," Proc. of 4th Inter. Conf. on Autonomic and Autonomous Sys., pp. 76-82, 2008.

[38] Y.C. Hsu, A.P. Chen, C.H. Wang, "A RFID-Enabled Traceability System for the Supply Chain of Live Fish," Automation and Logistics, pp. 81-86, 2008

[39] P. Jones, "Networked RFID for use in the Food Chain," Proc of the 2006 Emerging Technologies and Factory Automation Conference, pp. 1119-1124, 2006.

[40] F. Gandino, B. Montrucchio, M. Rebaudengo, and E.R. Sanchez, "On Improving Au-
      tomation by Integrating RFID in the Traceability Management of the Agri-Food Sec-
      tor," IEEE Transactions on Industrial Electronics, Vol. 56, Issue 7, pp. 2357-2365, 2009.

[41] L.C. Wang, Y.C. Lin, and P.H. Lin, "Dynamic mobile RFID-based supply chain con-
      trol and management system in construction," Advanced Engineering Informatics,
      Vol. 21, Issue 4, pp. 377-390, 2007.

[42] A.S. Martinez-Salaa, E. Egea-Lopez, F. Garcia-Sancheza and J. Garcia-Haroa, "Track-
      ing of Returnable Packaging and Transport Units with active RFID in the grocery
      supply chain," Computers in Industry, Vol. 60, Issue 3, pp. 161-171, 2009.

[43] M. Tu, J.H. Lin, R.S. Chen, K.Y. Chen, and J.S. Jwo, "Agent-Based Control Framework
      for Mass Customization Manufacturing With UHF RFID Technology," IEEE System
      Journal, Vol. 3, Issue 3, pp.343-359, 2009.

[44] K. Ohashi , S. Ota, L. Ohno-Machado, and H. Tanaka, "Comparison of RFID Systems
      for Tracking Clinical Interventions at the Bedside," Proc of AMIA Annu Symp 2008,
      pp. 525–529, 2008.

[45] Chung-Chih Lin , Ping-Yeh Lin, Po-Kuan Lu, Kuan-Yu Hsieh, Wei-Lun Lee, and
      Ren-Guey Lee, "A Healthcare Integration System for Disease Assessment and Safety
      Monitoring of Dementia Patient," IEEE Transactions on Information Technology in
      Biomedicine, Vol. 12, No. 5, pp. 579-586, 2008.

Chapter 4

Implementation of a Countermeasure to Relay Attacks for Contactless HF Systems

Pierre-Henri Thevenon and Olivier Savry

Additional information is available at the end of the chapter

1. Introduction

Nowadays, HF contactless technologies following the ISO 14443 standard are extensively used worldwide. Critical applications like access control or payment require high security guarantees. However, contactless channels are less secure and offer more opportunities for any kind of intrusion than other ways of communication; e.g. eavesdropping and contactless card activation using false reader [1, 2, 3, 11]. Among the attacks on the physical layer, relay attack is the most dangerous because of its simplicity, its impact and its insensitivity to cryptographic protections. It consists in setting up an unauthorized communication between two devices out of their operating range [4, 6]. On Figure 1, two attackers are able to create a link between the reader and the contactless card without the agreement of the owner. A relay is composed of two elements: a first one close to the reader and called proxy, a second one close to the card and called mole. These two elements communicate together by a wired or a wireless link

Figure 1. Relay scenario in a queue

A relay attack is thoroughly transparent for current contactless systems and cryptographic protocols. A possible countermeasure is the distance bounding protocol which can add an upper bound for the distance between the two communicating devices.

In this chapter, we will first assess the potential of relay attacks by implementing them and by keeping in mind the concern of introducing a delay as low as possible. Indeed, this time remains the only detectable feature of such an attack and the existing countermeasures rely on its accurate assessment.

The delay constraint guides us towards the development of three kinds of relay: a wired passive relay, a relay based on a wireless super-heterodyne system and a wireless relay with a complete demodulation of the signal. Our experimental results show that those cheap devices introduce really low delay from 300 ns to 2 μs jeopardizing the use of current distance bounding protocols. A more adapted solution will then be implemented and addressed in the second part of the document. It modifies the stage of the distance bounding protocol which uses the physical layer to carry out a delay assessment with a correlation in the reader between the received signal and the expected one. Finally, a security analysis will be performed and improvements will be discussed.

2. Relay attacks

2.1. Related work

The relay attack is based on the Grand Master Chess problem described by Conway in 1976. The latter shows how a person, who does not know the rules of this game, could win against one of two grand masters by challenging them in a same play. The relay attack is just an extension of this problem applied to the security field. By relaying information between a reader and a card outside the reader field, an attacker can circumvent the authentication protocol. This attack needs two devices: a mole and a proxy. The mole pretends to be the true reader and exchange data with the proxy which pretends to be the true card.

The larger the distance between the different elements is, the more efficient is the relay. Typical maximum distances between the reader and the proxy or between the mole and the card are roughly 50 cm. The distance between the mole and the proxy is not limited; it just depends on the chosen technology [5].

By using a relay, an attacker can transmit requests and answers between an honest reader and an honest card separated by 50 metres [6]. Many communication channels can be used to link the mole and the proxy like GSM, WIFI or Ethernet [8]. The delay, introduced by such a relay is more than 15 μs. At the physical layer, this attack is the most dangerous for many reasons:

- The card is activated and transmits information when it is powered, without the agreement of the victim. Anyone can be a victim because the attacker has just to be close enough to control your card like in a crowd.

- The attack occurs on the physical layer i.e. the relay transmits coded bits without knowledge about the frame significance. The ISO9798 standard presents an authentication protocol to prove that the contactless devices involved in the communication share the same secret key. For eavesdropping or skimming attacks, the use of this kind of protocol limits the risks. For the relays attacks, knowing the key is not necessary. Actually, a relay does not neither modify the information of the frame nor has to know its meaning. It just transmits the data. The encrypted data are transmitted as plain text.

- Contactless standards such as standard ISO14443 impose timing constraints in order to synchronize data sent by many cards at the same time, especially during the anti-collision protocol. However, these constraints are not enforced by the majority of cards [9]. These requirements would complicate the relay attack if they were really applied. Another weakness of the standard is the time delay between the reader request and the card answer. These time delay is not only such long but also expandable by the card and consequently by an attacker.

2.2. Presentation of relay attacks

The delay in current relays is mainly due to the use of components such as microcontrollers or RFID chips. This kind of components is used for the reconstruction of the decoded signals. So, the original signal becomes compatible with other protocols, like Wifi or GSM, used in the wireless communication between the mole and the proxy. All these signal processes lead to the addition of delays in the relay. They can be considerably lowered by the only use of analog components. Attack scenarios with wired relays must then be considered because they can induce very low delays. Moreover, this kind of relays is simple to realize, with few cheap components. Even if they seem to be unlikely, they can be effective in a queue for example or if they are hidden in the environment.

2.2.1. Passive wired relay

Fig. 2 depicts a simple design of a relay which introduces a very low delay close to a period of the carrier 13.56 MHz. This relay does not require an amplifier or other active components. The coaxial cable between the two antennas can be longer than 20m. Such a system is very low cost; the attacker needs a piece of PCB, few components for the matching and a coaxial wire. Overall cost is a few dollars at most. We claim that wired relays are the simplest and fastest relays by design and as a consequence, they should challenge the approaches of countermeasures which only parry the largest delays.

2.2.2. Relay based on a wireless super heterodyne system

This relay, shown on fig. 3, is quite similar to the relay attack developed by Hancke because it is not restricted by a wired link. Contrary to Hancke's relay, our wireless relay does not use digital components like microcontrollers or RFID chips to process the signal. The delay induced by this relay should be shorter. To do so, the reader signal of frequency fc is mixed with another signal of frequency F, generated by a local oscillator. It results a signal of frequency fc+F, easier

Figure 2. Potential use of a wired relay

to amplify and to send further. A PLL is used as a local oscillator to have the same frequency in the modulation and demodulation circuit.

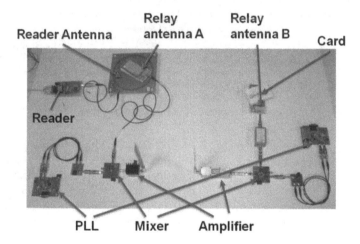

Figure 3. Forward wireless relay

2.2.3. Relay with demodulation of the signal

We have developed a more advanced relay (Fig. 4) close to those realized by Hancke or Kasper.. To realize a relay which demodulates the signal is more complex for an attacker, because it must have a perfect knowledge of the contactless standards. Our system is compliant with the

ISO14443-A standard to be compared with literature relays. However, it can be adapted to a different contactless standard such as ISO15693 or ISO14443-B.

The proxy is mainly based on thus developed by Carluccio et al. [7]. This electronic card can be divided into two subsystems: one for demodulation and decoding of the reader signal and one for the load modulation of the card.

The Mole is based on a reader developed in our laboratory. This device has a RF front-end RF which allows amplitude modulation and demodulation. The heart of the mole is a FPGA which separates the phase of emission and reception phase. The proxy signal is processed by the FPGA of the mole; it is coded in modified Miller and modulated in OOK. The HF signal is then amplified and injected in the antenna. The victim's card understands the request of our Mole as a frame from a standard reader and answers by modulating its load. This signal is firstly processed by an analog system and then sampled, demodulated and decoded by the FPGA.

The proxy and the mole communicate together through a wireless system. We have used chips used in the video/audio wireless transmission systems since they allow a sufficient bit rate of 212kbits/s.

The datasheet of video transmission systems provides a theoretical distance operation of 100 metres. In practice, problems of propagation in a building must be taken into account but this distance is sufficient to realize the attack in a shop. Based on experiments realized with the relay, we have obtained a maximum distance of 10 cm between the card and the mole but also between the proxy and the reader.

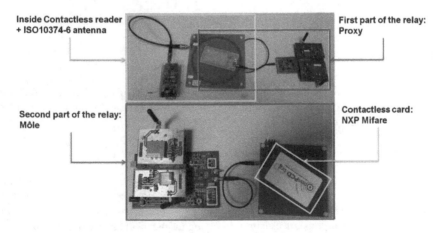

Figure 4. Relay with demodulation of the signal

2.3. Experiments on introduced delay

This experiment is performed to measure the introduced delays of the different relays. To do so, a reader sends a fixed sequence through a relay. With an oscilloscope, this sequence is recorded directly on two calibration coils located close to the two relay antennas. This sequence is a signal modulated in amplitude with a subcarrier at 848 kHz. The cross-correlation of the two recorded signals allows the computation of the temporal shift between them. In this experiment, we assume that the delay is the same for the forward and the backward channel so the results are the double of the value which is computed.

Fig. 5 gives an overview of the computed delays. Each type of relay is characterized by a temporal distribution. The delay introduced by the relay can be used to detect the presence of a relay. Wireless relays and wired relays have roughly the same delays because the mix of the signals is very fast in the case of the wireless relay. The relay with demodulation introduces a delay 7 times inferior to Hancke's relay.

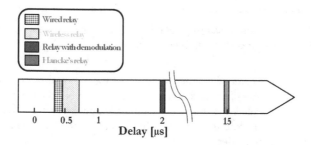

Figure 5. Measured delays

3. Countermeasures

In this part, we first describe the main existing relay detection systems and have a critical look to these solutions. Then, a new protocol based on the correlation, is described and implemented on a real contactless system.

3.1. Existent countermeasures and weaknesses

As mentioned before, design of a suitable countermeasure against relay attacks is a veritable challenge. This is partially because cryptography has no effect on it. Currently, there are few methods to detect relay attacks: distance bounding protocols, countermeasures based on timing measurements or physical structures implying the denial of service of the card.

3.1.1. Distance bounding protocols

In 2003, Hancke et al. have presented the first distance bounding protocol designed for contactless systems [11]; it is based on Brands and Chaums description [10]. Since then, many others distance bounding protocols have been published to improve the security of the scheme.

However, if they have been designed to use the physical layer of the system, they never have been implemented and tested in the HF band.

Distance bounding protocols are used to detect additional delays introduced by a relay during a transaction between two devices. This kind of protocol is often divided into three stages. In such a protocol, the card, named prover, must convince the reader, named verifier, that they are close to each other. During the first stage, the verifier and the prover exchange encrypted sequences, used during the second stage. While, the second stage consists of a timed exchange between the prover and the verifier, in order to verify the card's location. This analysis is made by measuring the time between the request of the verifier and the response of the prover. The last stage is an authentication and verification. The verifier computes and checks the measured times to define the location of the prover and analyses the prover answers to verify its honesty.

The reliability of such protocols depends mainly on the physical layer; the communication channel during the exchange stage affects the accuracy of the propagation time measurements. However, Hancke et al. and recently Rasmussen et al. are the only authors who gave a number of indications related to the protocol implementation at the physical layer level [13]. Other authors claim the merits of their distance bounding protocols such as cost, complexity, reliability but none of them has treated the problem of the protocol implementation for a contactless system.

Such discussions and analysis may be proposed before further works on these protocols.

Distance measurements based on the use of electromagnetic and acoustic waves are used in many applications such as radars. The distance resolution is inversely proportional to the bandwidth; this relation shows one of the weaknesses of distance bounding protocols implemented on a contactless or UWB communication channel:

These two communication channels use electromagnetic waves which have celerity close to the speed of light. In a contactless system, the distance between the verifier and the prover is smaller than 10 cm. Propagation time is then smaller than 300 ps. The first assumption of distance bounding protocol is that the processing time of the signal is assumed to be much smaller than the propagation time of the signal transmitted between the two parts, so smaller than 300 ps.

- HF communication channel: For a contactless system with a bandwidth of 848 kHz, the spatial resolution is around 350 m. Such resolution is too weak to measure a distance between two communicating entities. Moreover, establishing time in HF antennas, processing time for modulation and demodulation take too much time to measure small delays

- UWB communication channel: The bandwidth of a UWB system is equal to 20-25% of its central frequency. The spatial resolution is then close to 1.6 m for a 1GHz UWB system. Such resolution is suitable to detect any kind of relays. However, UWB implementation on an HF contactless system is complex. Hardware constraints are required such as the modification of all RF front-end: add of electrical antennas and specific modulation and demodulation systems.

To summarize, distance bounding protocols are really difficult to implement since the use of the UWB adds cost and complexity. By using HF communication channel, the propagation time remains difficult to isolate because it can be small compared to the processing time. To consider the constraints imposed by the physical layer of HF contactless systems is a priority for developers of algorithms against relay attacks

3.1.2. Solutions based on time measurements

Reid et al. have proposed a solution allowing the measurement of the time duration between the end of the request and the start of the reply [9]. For that, the authors have identified two reference points which represent the state change of the system. In theory, this system can measure average delays of 300 ns; this resolution is 50 times smaller than the delay introduced by Hancke's relay. This counter-measure can be accurate enough to avoid relay attack. However, some problems remain:

• The card does not always reply at the same time;

• No protocol authentication are implemented;

• The signal processing can increase the duration of the delay;

• The attacker can act on the relay to disturb the counter measure.

Munilla et al. have proposed a protocol based on the ISO14443-A standard [10]. In this solution, the reader measures the delay between its request and the card answer. It computes the number of carrier periods between the end of its synchronization bit and the time when the carrier becomes stable after the card response. The authors concluded that their protocol can be used to detect simple relay attacks which induced delays lower than 1 µs. However, this resolution is inefficient against distance fraud attack. Moreover, this countermeasure imposes the modification of standards and of the physical layer. In this solution, the carrier is switched off regularly so the card cannot be powered during this time.

3.1.3. Solution based on the denial of service

The literature provides some examples of solutions that enable the card's holder to disable their card temporarily [1, 18]. The easier solution is a wallet made of metallic sheets, which acts as a faraday cell. Reference [17] presents physical structures which enable the card's holder to turn off their card by separating the chip and the antenna.

3.2. Our solution

This part describes a new protocol compliant with contactless standards which authenticates the two communicating parties. A first implementation of this countermeasure on a real contactless system demonstrates its reliability.

3.2.1. The proposed scheme

The main objective of this countermeasure is to detect relay attacks by measuring the delay introduced by them by using the correlation method.

The first assumption of our protocol is not based on the propagation time but on the complete delay between a triggering pulse of the reader and the answer of the card received by the reader. This delay is different when a relay is inserted between the reader and the card. For an easier understanding of the solution, we suppose that the forward and backward times induced by the relay are the same to make the explanations easier. In this solution, a recorded sequence is correlated with the same sequence sent by the card. The solution is based on an authentication of the card and the measurement of delay induced by a potential relay, as shown in the fig. 6 and the fig. 7. Our protocol is similar to the distance bounding one; it could be divided into three stages: initialization, time measurement and verification.

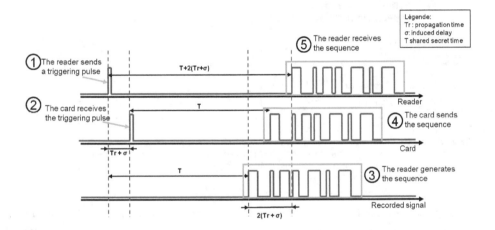

Figure 6. Time measurement stage in the proposed protocol

The first part of our protocol starts by the sending of a nonce from the reader to the card. The reader and the card use any validated symmetric lightweight cryptographic algorithm E, a shared key k and an exchanged random number to calculate T, the waiting time before the sending of the card answer and S, the sequence send by the card and synthesized in the reader. Hence, the computation of T and S by the reader and the card allows the authentication of the card. The first objective of our solution is to detect the relay so there is no mutual authentication in this protocol. However, few modifications of our protocol are possible to have this option.

After the exchange of the random sequence, the second stage starts (fig. 6). A random number of clock cycles after the end of the request frame, the reader modulates briefly its field to create a synchronization pulse. This pulse is received by the card with a delay function of the propagation time Tr and the delay induced by the uplink relay σ. It acts as a start point of the protocol for the reader and the card. Once the triggering pulse is received, the card has to send

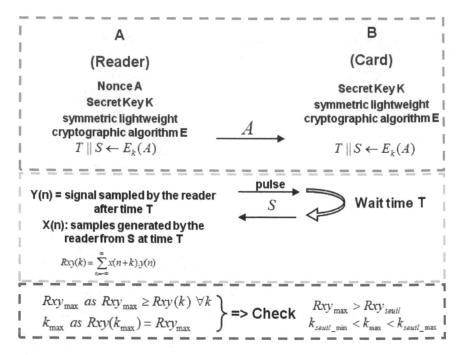

Figure 7. The proposed protocol consists of three stages. The first stage calculates two pseudorandom numbers, using a symmetric cryptographic algorithm and a secret key k and an exchanged nonce. The second stage is time critical as the card has to answer one of the generated pseudorandom sequences, a time T after a synchronization pulse. The third is the verification of the relay presence

the sequence S after a time duration T measured from the synchronization start by using its load modulation. The time duration between the reader request and the card answer is usually sufficient to send this sequence. The reader received the sequence S with a delay from its sent synchronization. This delay depends strongly on the delays introduced by the uplink and downlink relay. A time T after its synchronization pulse, the reader synthesizes the sequence S as it was sent by the card but without any delay. The received sequence S from the card is sampled by the reader after a time T from the triggering pulse to synchronize the samples Y(n) from the card answer and the sample X(n) from the synthesized sequence of the reader.

During the verification stage, the reader correlates the two recorded sequences X(n) and Y(n) to determine the delay between the two sequences. The index corresponding to the maximum value of the correlation is the number of samples of the delay. This number and the maximum correlation value are used to determinate the presence of a relay in the reader field.

3.2.2. Experiments and results

This part presents the first results of correlation based on our solution implementation. The solution is implemented on an "open" reader and contactless card that we developed, illustrated in fig. 8.

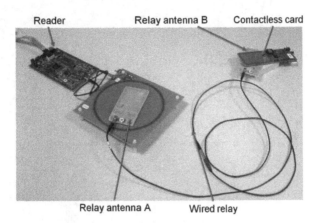

Reader   Relay antenna B   Contactless card

Relay antenna A   Wired relay

Figure 8. Experimental setup with an open reader and an open card in the presence of a wired relay.

The objective of our experiment is to demonstrate that the computed delay depends on the relay. We perform our four scenarios: one without relay, the others with the three relays studied previously.

Based on 2000 delay values taken in presence or not of a relay and for different distance between antennas, we compute the distribution of these four cases.

This first implementation of a cryptographic protocol based on the physical layer gives interesting results. The chart on fig. 9 shows three different histograms: one for each implemented relay (the wired relay and the fastest wireless relay) and one in the case without relay. These first results prove the efficiency of our solution since it is able to detect a relay with the help of the maximum delays occurred in a classical contactless system. Only relay designed by us are tested but we assume that delay induced by these relay are close to the theoretical minimal delay induced by the most critical relay. Then, we can claim that our solution can detect the most existed relay attacks.

3.3. Discussion

The objective of this discussion is the analysis of the security and the privacy of this solution.

Card cloning and replay attacks with a false card could not be authenticated by the reader and the threat will be detected. In fact, the card must compute two random binary sequences during the first stage of our protocol. The result of this computation is checked during the second

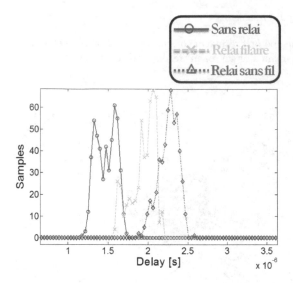

Figure 9. Delay distribution with our solution for each case (with one of the two most critical relays or without relays) for different distance between antennas.

stage. A false card could not send the correct sequence at the correct time to the reader because they depend on the knowledge of the secret key k.

In the case of distance bounding protocols, the security is analyzed by exposing the protocol to three different attacks. Our solution can be exposed to the same attacks to detect possible weaknesses.

3.3.1. Distance fraud

The scenario of this first attack requires a true reader, named verifier, and a false contactless card, named prover. The prover must convince the verifier they are close to each other when it is outside the communicating range. Firstly, this attack is only theoretical in the domain of contactless systems since no author implements this attack. Thus, the prover authenticates the card during the challenge; a corrupted card will be detected (see above). The detection of distance fraud attacks depends on the delays introduced by a modified card.

3.3.2. Mafia fraud

In the mafia fraud attack, the attacker does not perform any cryptographic operations based on the security protocol, and only forwards the challenges and the responses between the honest prover and the honest verifier: it is the standard relay attack.To convince the verifier and the prover they are close to each other, the relay can speed up the clock of the carrier to improve the response time of the prover answer [14]. The received signal will be compressed and the correlation value will be weaker so this attack will be detected. In the same way, the

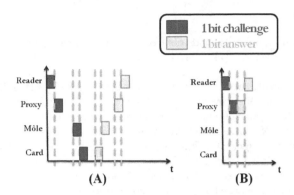

Figure 10. Noisy environment (A) classical case (B) Case with the anticipation of the answer

relay can not anticipate the synchronization pulse of the card because the pulse position in the time is random. Our protocol is resistant to the mafia fraud.

3.3.3. Terrorist fraud

This attack is similar to the previous one, the only difference is that the contactless card and the relay cooperate to mislead the reader. This attack is possible if the protocol does not guarantee a link between the authentication part and the timed challenge part. In our case, the answer of the card and the time between the pulse and this answer are deduced by the cryptographic key during the authentication part. Our solution is resistant to terrorist fraud.

3.4. Physical attacks

The main objective of this article is to prove the reliability of a solution based on the HF physical layer. We assume that the authentication protocol can be improved based on the literature. However, our solution must be resistant to such physical attacks.

3.4.1. Noisy environment

Distance bounding protocols are usually based on the use of an Ultra Wideband modulation. This modulation is sensitive to noise because its spectral power density is weak. In the case of a noisy channel, the attacker can anticipate the bits sent by the card to reduce the value of the delay measured by the reader. The answer of the card is just one bit; the attacker has a fifty-fifty chance to discover the real value. Then the reader can believe these errors are due to the noisy environment since they are introduced by the attacker. Then the reader concludes that the card is closer than it is and it does not detect the relay (fig. 10).

In our solution, the use of the HF physical layer which is less sensitive to noise and a length of many bits for the sequence S circumvent the anticipation of the sequence by the relay.

3.4.2. Timing attacks

The clock of the card is linked to the carrier frequency of the device which is powered it. This attack, described by Hancke [14], allows an attacker to speed up the clock and then the processes computed by the card to reduce the secret time T of our protocol. Then, the relay transmits the card answer earlier and the relay is not detected (fig. 11). In [16], the authors show that few solutions allow the limitation of the clock increase such as low-pass filters or internal clock. With this kind of solutions implemented on the card, an attacker can absorb 2-3 ns by clock cycle (73.74 ns). To realize such attacks, the attacker has to use a complex relay which demodulates the signal. This kind of systems introduces delays of few μs. Then this attack is not possible if the secret time T between the reader request and the card answer is lower than a determined threshold. This threshold must be inferior to the necessary time to compensate the delay introduced by the processing times of the relay.

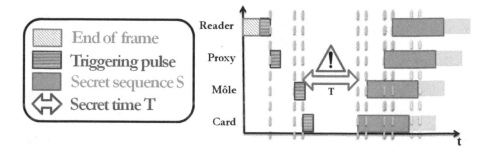

Figure 11. Timing attacks

3.4.3. Anticipation of the synchronisation pulse

The anticipation of the pulse by the relay is a weakness of this kind of protocols because our pulse does not contain a challenge. The relay does not have to wait the pulse and can anticipate and send it earlier. This solution cancels the delay introduced by the forward processing times of the relay (fig. 12). A first solution is to send the pulse just after the end of frame of the reader. Then, the attacker can just cancel the delay introduced by the forward relay. Secondly, our system can use multi level modulation to encrypt the pulse. This modulation can be in amplitude or phase. The value of the secret time T and the secret sequence S can be linked to the value of the modulation level.

Then, this solution limits the anticipation of the pulse since the answer of the card is function of the modulation of the reader.

3.5. Countermeasure improvement

The accuracy and the reliability of our solution can be enhanced:

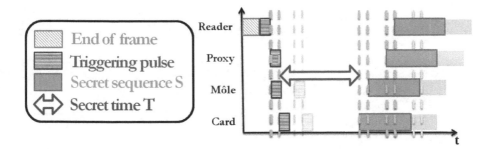

Figure 12. Anticipation of the synchronisation pulse

3.5.1. Pulse detection

An important improvement concerns the detection of our reference point; the accuracy is mainly due to the triggering pulse identification. It is currently realized using a binary signal, this signal results of the demodulation of the RFID signal. We do not control this demodulation but we suppose that it adds a shifting delay to our total delay. We have to develop a system which can detect a pulse with a fixed delay to reduce delay accuracy significantly. The improvement of the accuracy and the rapidity of the pulse detection can be made by using phase modulation only for the pulse. This solution has been implemented on the previously used contactless reader and a new contactless card able to decode a signal modulated in phase. Our approach, c.f. B.1?, was tested with the new parameters for the pulse emission and reception. The results are described on fig. 13. The delay distribution for the case without relay and the case wired relay show an important improvement. Indeed, the two histograms are significantly different; the introduced delay becomes more important with the presence of the wired relay. This experiment shows that all relay attacks can be detected efficiently using the phase modulation for the synchronization pulse. However, this improvement implies the modification of the existing Radio-Frequency front-end equipment.

3.5.2. M-sequences

M-sequences present many properties which can improve the accuracy and the sequence generation of our solution. An M-sequence is a pseudo random sequence generated in most cases by linear feedback shift register and is used in many cryptographic applications. Two properties of M-sequences are of interest: randomness and correlation properties. The sequence is composed of pulses with variable width multiple of the minimal period. The autocorrelation of this kind of signals is an approximation of a Kronecker delta function. Such functions present an important peak when there is no delay between the signals is null which is easy to detect in the case of an implementation.

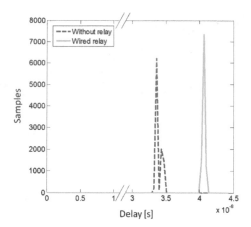

Figure 13. Delay distribution with our solution for the case without relay and in the case with a wired relay for different distance between antennas.

3.5.3. Correlation on PM (Phase Modulation) signals

In the case of NFC use in smartphones for critical application, we can suppose that the target (corresponding to the contactless card) uses active mode to answer to the initiator (the reader in our solution).

Then, the target can modulate its signal by varying the instantaneous phase of the carrier signal. The phase modulation can be more complex for implementation but more accurate in terms of correlation. In fact, the signal received and recorded by the initiator must be in phase with the generated one. There are fewer problems with establishing times in antennas because there is no subcarrier, c.f. II.C.2. The obtained accuracy depends on the phase modulation but we can think that we can detect delays close to half of a carrier. Such improvements imply modifications of standards.

4. Conclusion

The relay attack is an attack on physical layer which should be seriously considered because it can be easily implemented and used in a lot of applications. Moreover, the increasingly use of NFC technology, especially in phone applications, opens new opportunities for intruders. Nowadays, contactless readers are unable to detect a relay. This attack does not modify the signal, nor disturb the transaction and induce delays close to a few periods of the signal carrier. Additionally, cryptography, which is the best solution for most threats, cannot detect this attack.

The first objective of our work was to realize relay attacks with the shortest delays. Within this chapter, we have presented three different solutions to overcome this problem. Experiment results show that the designed wired relay is the most critical relay in terms of the introduced time delay. Our work shows that with two simple antennas and a wire, an attacker can relay data between a reader and a card with delays close to 300 ns, i.e. 50 times shorter than Hancke's relay attack. Today, no countermeasure is able to detect this kind of relays.

The second objective was to develop a new solution to detect such delays with maximum certainty. This countermeasure uses correlation between two sequences to compute the delay introduced by the relay. This will be used to determine the presence of a relay in the reader's field. For the first time, a solution was implemented on a contactless system and the results are interesting. A contactless system does not require additional hardware resources to use our protocol which allows accuracy close to 300 ns. This solution respects the contactless standards and does not disturb the communication between the reader and the card since the protocol can run during the response time of the card. Apart from the most critical relay, namely wired relay, which is not detected in few rare cases, all kind of relays are detected with our counter-measure. However, we developed another solution that detects all kind of relays attacks by improving the accuracy of our contactless system. However, the latter requires a modification of the RF front end.

Author details

Pierre-Henri Thevenon and Olivier Savry

Léti, Minatec, CEA Grenoble, France

References

[1] Juels A. RFID security and privacy: A research survey, IEEE Journal on Selected Areas in Communications, Vol. 24, Issue 2, p381–394; 2006.

[2] Weis S., Sarma S., Rivest R., Engels D. Security and Privacy Aspects of Low-Cost Radio Frequency Identification Systems, proceedings of the International Conference on Security in Pervasive Computing, Vol. 2802, p454-469, SPC 2003; 2003.

[3] Hancke G.: Practical attacks on proximity identification systems, IEEE Symposium on Security and Privacy, p328-333; 2006.

[4] Hancke G., Mayes K., Markantonakis K. Confidence in Smart Token Proximity: Relay Attacks Revisited, Elsevier Computers & Security, Vol. 28, Issue 7, p615-627; 2009.

[5] Lishoy F., Hancke G. P., Mayes K., Markantonakis K. Practical NFC Peer-to-Peer Relay Attack using Mobile Phones, Workshop on RFID Security, RFIDSec'10, 7-9 June 2010, Istanbul, Turkey; 2010.

[6]   Hancke G. A Practical Relay Attack on ISO 14443 Proximity Cards, Manuscript; 2005.

[7]   Carluccio D., Kasper T., Paar C. Implementation details of a multipurpose ISO 14443 RFID-tool, Workshop on RFID Security, RFIDsec'06, 12-14 July 2006, Graz, Austria; 2006.

[8]   Kfir Z., Wool A. Picking virtual pockets using relay attacks on contactless smartcard systems, SecureComm 2005, 5-9 September 2005, Athens, Greece; 2005.

[9]   Hlavac M., Rosa T. A Note on the Relay Attacks on e-passports: The Case of Czech e-passports, IACR ePrint; 2007.

[10] Brands S., Chaum D. Distance Bounding Protocols, Advances in Cryptology, p344–359, Workshop on the Theory and Application of Cryptographic Techniques, EURO-CYPT'93, May 23-27, 1993, Lofthus, Norway; 1993.

[11] Hancke G. Eavesdropping Attacks on High-Frequency RFID Tokens, Workshop on RFID Security, RFIDSec'08, Budapest, Hungary; 2008.

[12] G. Hancke, M. Kuhn, An RFID distance bounding protocol, SecureComm 2005, 5-9 September 2005, Athens, Greece; 2005.

[13] Rasmussen K. B., Capkun S. Realization of RF Distance Bounding, 19th USENIX Security Symposium, USENIX'10, 11-13 August 2010, Washington, DC, USA; 2010.

[14] Hancke G., Kuhn M. Attacks on Time-of-Flight Distance Bounding Channels: roceedings of the first ACM Conference on Wireless Network Security, WiSec'08, p194–202, 31 March – 2 April 2008, New York, USA; 2008.

[15] Munilla J., Ortiz A., Peinado A. Distance Bounding Protocols with void-challenges for RFIDs, Workshop on RFID Security, RFIDsec'06, 12-14 July 2006, Graz, Austria; 2006.

[16] Reid J., Gonzalez Neito J., Tang T., Senadji B. Detecting Relay Attacks with Timing Based Protocols, ACM Symposium on Information, Computer and Communications Security, ASIACCS 2007, 20-22 March 2007, Singapore; 2007.

[17] Karjoth G., Moskowitz P. Disabling RFID Tags with Visible Confirmation: Clipped Tags Are Silenced, Workshop on Privacy in the Electronic Society, WPES'05, 7 November 2005, Alexandria, VA, USA; 2005.

[18] The off switch for "always on" mobile wireless devices, spychips, toll tags, RFID tags and technologies. www.mobilecloak.com (accessed 12 September 2012).

[19] DIFRwear's RFID Blocking Products. www.dirfwear.com (accessed 12 September 2012).

Chapter 5

Localizing with Passive UHF RFID Tags Using Wideband Signals

Andreas Loeffler and Heinz Gerhaeuser

Additional information is available at the end of the chapter

1. Introduction

Localization of positions and detection of objects is a key aspect in today's applications and, although the topic exists a while ago, it is still under ongoing research. The introduction of global navigation satellite systems (GNSS), particularly GPS [1], and its improvements with accuracies down to a few meters, was a huge step towards ubiquitous localization [2, 3]. This is almost valid for outdoor environments, whereas indoor localization is still a challenging issue [4, 5]. The reason for that is the demanding, dynamic indoor environment, causing severe multipath fading, leading to hard predictable propagation models - thus influencing time, power and phase measurements. However, in the past, much effort has been put into designing high accurate indoor localization systems, including technologies like ultrasonic sound, infrared light, Wi-Fi, Bluetooth, ZigBee, cellular mobile communication (GSM, UMTS), ultra-wideband and RFID just to mention a few of them. Despite all the effort, there is no outstanding technology comprising all indoor localization contingencies as every technology in use has its advantages and disadvantages regarding accuracy, availability, complexity and costs.

Due to constantly falling prices of UHF RFID tags [6] additional applications arose beside the traditional concept of radio frequency identification (RFID). Major applications include supply chain technologies [7] and logistics [8], from container level tagging even down to item level tagging [9]. Regarding the Internet of Things [10], UHF RFID has some advantages over other RFID technologies, i.e., LF and HF: UHF RFID tags are small, do not require a battery, allow high data rates and high reading ranges, whereas LF and HF cannot serve with these issues at the same time [10]. Together with the mentioned low costs, the UHF RFID technology may be available in lots of objects (walls, carpets, doors, etc.) in the future. Therefore, indoor positioning using UHF RFID technology could be one solution towards ubiquitous localization, as efforts are made to shrink the size of RFID reader ICs and to integrate them into mobile phones.

The chapter is organized as follows. Section 2 gives a brief overview of today's wireless position-ing technologies with a focus on RFID. Section 3 introduces the proposed positioning system and shows the theoretical approach along with an example. Section 4 focuses on challenges and limitations of the system and Section 5 presents results from measurements carried out underlin-ing the principle of operation. Section 6 provides a discussion based on the results. Finally, Section 7 gives a short summary and concludes with a perspective for future work.

2. Basics and state of the art

This section provides an overview of state-of-the-art wireless positioning technologies. The section is divided into two subsections, with the first subsection describing measurement principles for positioning, whereas the second subsection has a focus on current positioning technologies based on RFID, particularly UHF RFID within the 900 MHz frequency band.

2.1. Positioning measurement principles

The first paragraph provides definitions for the terms precision, rightness and accuracy, whereas the following paragraphs describe the main positioning processes comprising lateration, angulation and fingerprinting. The last paragraph depicts the measurement techniques used for the positioning process, for instance, time of arrival, angle of arrival and received signal strength.

2.1.1. Precision, rightness and accuracy

Often, the terms "precision" and "accuracy" are used to define the same issue, namely how well a localization system or method works, e.g., the measurement error expressed in meters. However, precision and accuracy are not similar to each other. Therefore, this paragraph points out the differences and relations of the terms precision, rightness and accuracy.

Precision shows how well independent measurement values are located to each other. That means, if many measurement values are in dense proximity to each other, the measurement has a high precision; on the other hand, it does not mean that the measurement is accurate in any case. A standard term that is used to measure the precision is the standard deviation σ_x with

$$\sigma_x = \sqrt{E\left\{\left(\hat{x} - E\{\hat{x}\}\right)^2\right\}} \text{ and } \hat{\sigma}_x = \sqrt{\frac{1}{N-1}\sum_{k=1}^{N}\left(\hat{x}_k - \bar{\hat{x}}\right)^2}, \text{ with} \tag{1}$$

$$\bar{\hat{x}} = \frac{1}{N}\sum_{k=1}^{N}\hat{x}_k \tag{2}$$

$\hat{\sigma}_x$ describes the estimated standard deviation of the measurement, N describes the number of measurements, $\hat{x}_k$ the measurement value at the kth measurement, $\bar{\hat{x}}$ the estimated mean value of the measurement values. $\hat{x}$ describes the random variable of the measurement process, whereas E{·} is the corresponding expectation value. In the following, the standard deviation σ_x is used as a measure for the precision of a positioning technique.

Rightness or trueness describes how well the measured values respectively the expectation of the estimated values $\hat{x}$ fit to the expectation of the true values x, i.e., a so called bias with

$$Bias = E\{\hat{x} - x\} = E\{\hat{x}\} - x \text{ and } \widehat{Bias} = \bar{\hat{x}} - \bar{x} \tag{3}$$

$\widehat{Bias}$ is the estimated rightness of the measurement and $\bar{x}$ is the mean value of the true values. The rightness is a measure for the average discrepancy between a measured and a reference value and may be described as bias or offset.

Accuracy takes both, the precision and the rightness, into account. In fact, only high accuracy may be achieved if precision and rightness is high, too. A well known definition of the accuracy is the root mean square error RMSE, which is defined as

$$\text{RMSE} = \sqrt{\text{MSE}} = \sqrt{E\{(\hat{x} - x)^2\}} \text{ and } \widehat{\text{RMSE}} = \sqrt{\frac{1}{N}\sum_{k=1}^{N}(\hat{x}_k - x_k)^2} \tag{4}$$

$\widehat{\text{RMSE}}$ describes the estimated RMSE of the measurement and x_k the true value at the the kth measurement.

According to [11] the first expression in Equation (4) can be transformed into

$$\text{RMSE} = \sqrt{\sigma_x^2 + \text{Bias}^2} \tag{5}$$

Equation (5) shows that a distorted measurement with a high precision may be more accurate than an undistorted measurement with a low precision respectively standard deviation.

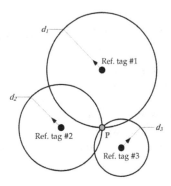

Figure 1. Example of trilateration with RFID reference tags

2.1.2. Lateration

Lateration is used to determine the position using distances to known reference points. For instance, an RFID reader may localize itself by evaluating distances to certain reference points, e.g., RFID tags, using the principle of trilateration, as shown in Figure 1. In this figure, two-dimensional (2D) positioning of **P**, an RFID reader, can be realized using three reference points, here reference tags. Assuming the reader is able to exactly determine its distance $d_i \; \forall \, i \in \{1,2,\,3\}$ to each of the tags, a circle is drawn around each tag with radius equal to the measured distance d_i. The intercept point of the three circles with radii $d_1 \ldots d_3$ indicates the position of the reader **P**. If the positions of the reference tags are known, the reader may determine its position by solving the set of equations

$$\sqrt{(x_P - x_i)^2 + (y_P - y_i)^2} = d_i, \quad i \in \{1,2,\,3\}. \tag{6}$$

$(x_P; y_P)$ is the position of the reader, which shall be estimated and $(x_i; y_i) \; \forall \, i \in \{1,2,\,3\}$ is the position of each of the reference points respectively tags. Solving the set of equations in (6) for three reference points yields [12, 13]:

$$\begin{pmatrix} x_P \\ y_P \end{pmatrix} = \begin{pmatrix} a_{1,2} & b_{1,2} \\ a_{1,3} & b_{1,3} \end{pmatrix}^{-1} \begin{pmatrix} g_{1,2} \\ g_{1,3} \end{pmatrix}, \tag{7}$$

with

$$a_{1,i} = 2(x_i - x_1), \quad i \in \{2,3\} \tag{8}$$

$$b_{1,i} = 2(y_i - y_1), \quad i \in \{2,3\} \tag{9}$$

and

$$g_{1,i} = d_1^2 - d_i^2 - (x_1^2 + y_1^2) - (x_i^2 + y_i^2), \quad i \in \{2,3\}. \tag{10}$$

In the case of three-dimensional (3D) positioning, a minimum of four reference points is necessary to unambiguously determine the exact position. However, due to the imperfectness of the distance measurement (noise, fading channel, etc.), there is usually no exact interception point, but rather an intersection area. Therefore, different error-minimizing algorithms can be used to make a best estimate for the position determination [14]. The accuracy of the measurements can be further increased by making use of more than the necessary minimum of reference points [15].

In RFID, generally, there exists clock synchronization between transmitter and receiver, as both components are located within the RFID reader. If, however, there is no clock synchronization between transmitter and receiver, the clock offset τ_{offset} will lead to a constant distance error d_{offset} within each range measurement. This additional parameter can be solved by adding one more equation (equal to one additional tag) to the minimum number of equations when there is no synchronization error:

$$\sqrt{(x_P - x_i)^2 + (y_P - y_i)^2} + d_{offset} = d_i \quad \forall\, i \in \{1,2,\, 3,4\} \tag{11}$$

As mentioned before, there should be no time offset in RFID systems. Nevertheless, constant phase shifts due to the non-constant reflection coefficient of RFID tags [16] can lead to an additional offset distance d_{offset}, having the same effect as a time-based clock offset. The set of equations in (11) describe hyperbolas rather than circles around the reference points. Figure 2 shows the effect of an offset distance d_{offset} and two out of four hyperbolic curves, which would intercept in position P.

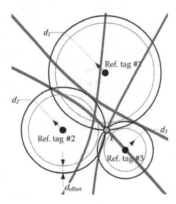

Figure 2. Example of hyperbolic lateration using RFID tags as reference points

2.1.3. Angulation

The principle of angulation rests upon the relations between angles and distances within a triangle; therefore, it is mostly common under the term triangulation. If two angles and one side of a triangle are known the remaining distances respectively the position to be determined can be calculated using the *law of sines* and the *angle sum of a triangle*. Figure 3 shows the principle used: Two antennas (Ant. #1 and Ant. #2) of an RFID reader are deployed to calculate the position of the RFID tag. This can be realized using, for instance, phase-based or direction-defined measurements. From independent angle measurements one obtains the angles α and β; the distance d_0 is known in advance. Subsequently, the remaining angle γ is calculated (angle sum in triangle) and from that the missing two distances d_1 and d_2 from the antennas to the RFID tag (law of sines). Angulation may be used in 2D or 3D localization problems.

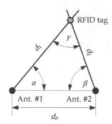

Figure 3. Triangulation example using two antennas to determine the position of an RFID tag

2.1.4. Scene-based localization / fingerprinting

Scene-based localization is divided in two sequential processes, a calibration process and an operational process. The calibration process records any environmental values (optical, electrical, physical, etc.), also known as fingerprints, at several positions within a scene and stores the data in a database [17, 18]. The following operational process is thus able to determine the position by measuring the current environmental values and comparing them with the values in the database. Special algorithms estimate the position by finding the position with the minimal error [19]. Figure 4 shows a room map with different WLAN base stations showing the electrical field strength at different locations [20] used along with WLAN positioning.

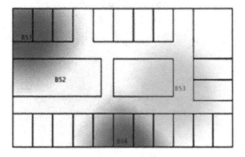

Figure 4. Electrical field strength distribution within a building to be used for WLAN positioning [20]

2.1.5. Positioning measurement techniques

After highlighting the measurement principles, this paragraph gives a brief overview over the common technologies used. Distance measurements may be based on measuring the time of flight, the signal strength and the phase between transmitted and received signal.

Time measurements include Time of Arrival (ToA) and Time Difference of Arrival (TDoA) measurements.

ToA measurements directly determine the distance by using the time of flight t_{ToA} of the signal. Multiplied with the corresponding propagation speed c, the speed of light in case of electromagnetic waves, this results directly in the distance d_{ToA} between transmitter and receiver as described in Equation (12). ToA measurements can be used directly along with trilateration methods.

$$d_{\text{ToA}} = t_{\text{ToA}} \cdot c \tag{12}$$

TDoA measurements determine the time difference of a signal received at known reference points rather than measuring directly the time between transmitter and receiver. This means, that the time stamp of the signal transmitted via the object to be localized is unknown, but the time differences at the synchronized receivers are determined. In contrast to ToA, TDoA does not require any synchronization between transmitter and receiver. The reference stations must be synchronized, indeed. One positioning method using TDoA measurements is hyperbolic lateration (see Paragraph 2.1.2).

RSS (Received Signal Strength) measurements are based on the received signal strength at the receiver. Hence, there are two possible candidates to process RSS-based data. The first one is based on the propagation conditions, usually including a modified and enhanced form of Friis transmission equation

$$P_r = P_t + G_t + G_r + 20\log\left(\frac{\lambda}{4\pi d}\right) [\text{dB}], \tag{13}$$

e.g., the log-distance path loss model

$$\text{PL}(d) = \text{PL}(d_0) + 10\alpha\log\left(\frac{d}{d_0}\right) + X \ [\text{dB}]. \tag{14}$$

Equation (13) describes the free space attenuation formula depending on the distance d and wavelength λ with receiving power P_r, transmitting power P_t, receiving and transmitting antenna gains G_t and G_r together with the free space attenuation $\left(\frac{\lambda}{4\pi d}\right)^2$ in dB. Equation (14) on the other hand describes the path loss $\text{PL}(d)$ depending on the distance d related to a reference path loss $\text{PL}(d_0)$ at distance d_0. The path loss may be described as the difference of transmitted and received power in dB. α represents the path loss exponent that depends on the propagation environment, whereas X is a zero-mean Gaussian distributed random variable describing the fading effects at different locations and instants of time. If, in case of the usage of Equation (14), $\text{PL}(d_0)$, α and the variance of X is known, one can calculate directly the probability for a certain distance d between transmitter and receiver. One disadvantage is that α and X are very dependent on the environment and can change significantly. The RSS measurements can be used along with lateration methods.

The second RSS-based approach is to measure in advance RSS values at certain positions within the localization area (fingerprints). The measured values are pre-processed and stored into a database. During the proper localization process, the current measurement values are compared to the values in the database and a best-fit position, based on the current values, is estimated. The advantage of using RSS values for this approach is that almost all devices come along with some kind of RSS-based output, including RFID readers. This method is used in scene-based positioning techniques.

Phase measurements can be used to provide information about speed, distance and angle. A good overview over these techniques is given in [21]. The radial velocity v of a tag is measured by evaluating the phase shift $\partial \varphi$ during different moments in time ∂t as given in Equation (15).

$$v = - \frac{c}{2\omega_0} \frac{\partial \varphi}{\partial t} \tag{15}$$

with c being the propagation speed and ω_0 the fixed circular frequency. The distance d between a tag and a reader can be calculated according to Equation (16) by measuring the phase shift at different frequencies.

$$d = - \frac{c}{4\pi} \frac{\partial \varphi}{\partial f} \tag{16}$$

Finally, phase measurements may be used to measure the angle θ between reader and tag (Angle of Arrival, AoA) using multiple receiving antennas. For two receiving antennas, Equation (17) describes the relation between the incoming angle θ, the phase difference $\varphi_2 - \varphi_1$ at a certain carrier frequency, and the spacing a between the receiving antennas.

$$\theta \approx \sin^{-1} \left[- \frac{c}{\omega} \frac{\varphi_2 - \varphi_1}{a} \right] \tag{17}$$

Phase measurement are used along with lateration and angulation principles to calculate the distance between transmitter and receiver respectively reader and tag.

2.2. Survey on UHF RFID-based localization systems

The following paragraphs provide a brief survey on state-of-the-art RFID localization systems within the UHF and microwave frequency band. The survey includes systems using RSS values, ToA and TDoA measurements, phase-based measurements as well as fingerprinting methods. Further surveys are provided in [22, 23, 24].

2.2.1. RSS-based direct range estimation

The SpotON system [25] is based on active RFID tags (working at 916.5 MHz) and provides a 3D ad hoc localization. RFID readers measure the signal strength of active RFID tags and a central server performs the calculation of the position within the environment. The relation between the RSS value and the position is based on the indoor channel model from Seidel and Rappaport [26]. The accuracy of the SpotON system is given with a cube of 3 m edge length, but this is dependent on the number of reference tags used. A disadvantage of the system is the long position calculation time from 10 to 20 s; an advantage is the easy to extend infrastructure and low system costs.

2.2.2. ToA-based range estimation

A 2.4 GHz RFID system based on SAW transponders is described in [27]. The SAW tags use a bandwidth of 40 MHz and reduce the echoes from the environment as the reflected tag signal is delayed due to the lower surface speed on the SAW material. The signal time on the SAW transponder is $T_{SAW}=2.2$ µs; so the reflections and echoes from the reader are almost faded out before the SAW-reflected signal responses back to the reader. A three-antenna system is used to perform a 2D positioning. However, the localization accuracy is strongly temperature-dependent and adds up to around 20 cm in a room with the dimension 2 m × 2 m.

2.2.3. TDoA-based range estimation

A localization system in the 5.8 GHz frequency band is described in [28]. The system is build upon active transponders and multiple base stations. One reference transponder is used as wireless synchronization source for the base stations. The system operates on the FMCW (frequency modulated continuous wave) principle (see [29]) and evaluates the time difference of a measurement transponder signal to determine the position of the measurement transponder. The position accuracy is given with 10 cm on an area of 500 m × 500 m.

2.2.4. Phase-based range estimation

The principle of FMCW is used to measure the distance to a certain object. The idea behind FMCW is to sweep a frequency band with the sweep rate α and record the phase and frequency differences. Furthermore, the transmitted signal from the reader is modulated by the transponder with a modulation frequency f_{mod}. The usage of a modulation frequency shifts the measurement signal into a higher frequency band (by f_{mod}), in order to suppress certain disturbances and noise within the baseband. The distance d is calculated through the frequency difference Δf and the phase difference $\Delta \varphi$ [30], with the latter providing a high range resolution within half a wavelength of the signal. Therefore, Δf provides a coarse distance estimation and $\Delta \varphi$ a more accurate one. $\Delta \varphi$ alone cannot be used as direct distance estimation due to ambiguities of the phase information. According to [30] the distance to a transponder can be calculated with

$$d_{coarse} = \frac{\pi \cdot c \cdot \Delta f}{2 \cdot \alpha} \text{ and } d_{precise} = \frac{c \cdot \Delta \varphi}{4 \cdot \omega_0}. \tag{18}$$

[31] describes an FMCW-based RFID system using a transponder with an UHF front-end working at 868 MHz. The transponder IC provides a modulation frequency of

f_{mod}=300 kHz and is driven by a 2.45 GHz FMCW signal with a bandwidth of 75 MHz. The system is tested on a cable-based setup and delivers an RMSE of 1 cm with cable lengths between 1 m and 9.5 m.

The system in [32] uses the phase difference observed at different frequencies to estimate the range between transponder and reader. The range estimation is performed according to Equation (16), whereas the maximum range d_{max} due to phase ambiguities is given with

$$d_{max} = \frac{c}{2B}.$$ (19)

However, the choice of the bandwidth B strongly influences the system's capabilities. A high B generates a high accuracy but a low maximum range; a low B leads to a higher range but at the expenses of a lower accuracy. Simulations at an SNR of 10 dB results in errors of 2.5 m for a frequency separation of B=1 MHz, and errors of 0.1 m for a B of 26 MHz. One has to keep in mind that the separation of 26 MHz is only valid within the US frequency band for RFID that ranges from around 902 MHz to 928 MHz. The European band is smaller (865.6 MHz to 867.6 MHz) leading to a lower accuracy.

2.2.5. Scene-based range estimation

LANDMARC [33] is an extension and improvement of the SpotON system [25, 34]. The system consists of fixed RFID readers, active reference tags (landmarks) and tags to be localized. The system uses RSS values connected with the kNN (k-nearest neighbor) algorithm [35] to estimate the position. The average error of the system is given with 1 m [33].

[36] examines the localization error of the LANDMARC system using passive, instead of active RFID tags. As a result, the orientation of the tags has a major influence on the total performance of the system. Using the kNN algorithm, in 47.5 % of the cases, the error was less than 0.5 m and in 27.5 % of the cases, the error was less than 0.3 m. However, in comparison to the original LANDMARC system, the overall range is smaller due to the usage of passive RFID technology.

A system based on a particle filter is proposed in [37]. It uses two RFID readers mounted on a small mobile vehicle to localize itself using RSS values. The calibration phase is performed in a room of size 5 m × 10 m. Depending on the speed of the vehicle and the material on which the tags are located (plastics, concrete, metal) the average error is between 1.35 cm and 2.48 cm. This system is based on the mobile robot system in [38] that incorporates a SLAM algorithm [39] based on Monte Carlo methods [40].

[41] describes a positioning system using fingerprints (RSS values and read rate) to localize tagged objects. First, a rough positioning is done using antenna cells, with each antenna illuminating a different room zone. This rough classification is realized using either Bayesian filter, kernel density estimation (KDE) based measurement models, support vector machines (SVM) or LogitBoost [42]. RSS-based values and read rate is used along with the algorithms to roughly estimate the position of the tagged object. One result was that the estimations based on RSS values perform better than the estimations based on the read rate. An even more

accurate positioning is realized when RSS values are used along with read rates of the transponders. Within the calibration phase, one tries to generate a high amount of reference points (fingerprints). Two algorithms are used and compared to perform within the positioning phase, a cascaded algorithm and a kNN algorithm. The cascaded algorithm runs the rough localization followed by the kNN algorithm for the high accuracy. The second algorithm resigns to use the rough position estimation. Similarly, the RSS-based fingerprints perform better than the read rates. Dependent on the environment, positioning errors between 37.9 cm and 42.1 cm may be achieved.

3. Wideband UHF RFID positioning system

This section introduces a brief motivation for the realized RFID positioning system before describing the basic structure of the system.

As derived from Section 2, current passive RFID localization systems suffer either from a high effort in the calibration phase (fingerprinting) or from bandwidth limitations which hold down the system's overall accuracy. Higher accuracies may be achieved using phase-based approaches at the expense of more complex hardware structures and necessary volume (see, for instance, phased array antennas [43]), only usable for fixed reader hardware. Therefore, an ideal passive mobile RFID positioning system should have:

- no change in hardware,

- high bandwidth,

- direct position estimation.

The here proposed system offers high bandwidth, but with very low power, and is based on a ToA method performing direct position estimation. As a consequence, additional hardware effort is necessary to provide the generation and evaluation of the high bandwidth signals.

In the following, a brief overview of the system, particularly its principle working structure, is provided.

Assuming a scenario as given in Figure 5. The scenario consists of n tags, whereby the distance to the ith tag has to be evaluated. The RFID reader is indicated at the bottom (only the coupler with antenna in monostatic mode) with input signal x_{reader} (into the antenna) and output signal y_{reader} (from the antenna). s_1 to s_n describe the backscatter modulation factors of the transponders, i.e., the factor with which the incoming signal from the reader is reflected with (principle of backscatter). If this factor is one, the complete signal is backscattered to the reader. Indeed, data from tag to reader is transmitted by varying this factor in time with the data to be sent [10, 44]. h_1 to h_n describe the bidirectional channel impulse responses between reader and tags. For reasons of simplification the following equations and terms are written without using the time t, although the expressions depend on it.

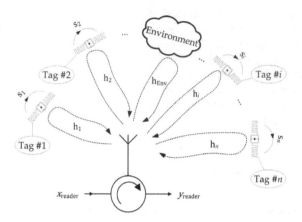

Figure 5. Scenario with passive RFID tags and reader in monostatic antenna setup

According to Figure 5 we can state (in time domain) by using the convolution $*$:

$$y_{\text{reader}} = [(h_1 * s_1) + (h_2 * s_2) + \ldots + (h_i * s_i) + \ldots + (h_n * s_n) + h_{\text{Env}}] * x_{\text{reader}} \tag{20}$$

For simplicity, let us assume that each backscatter modulation factor s_i has two modulation states according to

$$s_i = \begin{cases} s_{i,1} & \text{for modulation state 1} \\ s_{i,2} & \text{for modulation state 2} \end{cases} \tag{21}$$

In a first attempt, all tags are set into modulation state 1. The resulting signal $y_{\text{reader},1}$ is

$$y_{\text{reader},1} = [(h_1 * s_{1,1}) + (h_2 * s_{2,1}) + \ldots + (h_i * s_{i,1}) + \ldots + (h_n * s_{n,1}) + h_{\text{Env}}] * x_{\text{reader}} \tag{22}$$

In a second attempt, only tag #*i* is set into modulation state 2, all other tags stay in modulation state 1. This can be described as one small sequence of data transmission from the *i*th tag to the reader (uplink). The resulting signal $y_{\text{reader},2}$ is

$$y_{\text{reader},2} = [(h_1 * s_{1,1}) + (h_2 * s_{2,1}) + \ldots + (h_i * s_{i,2}) + \ldots + (h_n * s_{n,1}) + h_{\text{Env}}] * x_{\text{reader}} \tag{23}$$

The difference Δy_{reader} between both resulting signals $y_{\text{reader},1}$ and $y_{\text{reader},2}$ is

$$\Delta y_{\text{reader}} = y_{\text{reader},2} - y_{\text{reader},1} = [(h_i * s_{i,2}) - (h_i * s_{i,1})] * x_{\text{reader}} = [s_{i,2} - s_{i,1}] * h_i * x_{\text{reader}} \tag{24}$$

thus, taking the difference results into observation of the ith tag with the ith channel. By assuming that the tag's data contain the position of the tag (i.e., a reference tag), the reader has to evaluate the ith channel, regarding the range, to estimate the distance between reader and ith tag. In a 2D scenario, three tags must be read to get the position data, and three channels to the tags must be evaluated in order to localize the reader itself. The principle is described in more detail in Section 4.

The experimental hardware architecture of the reader is shown in Figure 6.

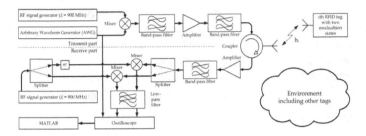

Figure 6. Experimental hardware architecture of the realized RFID localization system based on passive UHF RFID tags

The frequency-coupled RF signal generators generate carrier signals at the center frequency of f_c=900 MHz. The arbitrary waveform generator (AWG) creates the localization signal x_{reader} (in baseband). After upconversion of the localization signal, it is filtered, amplified and emitted into the environment through the antenna. The backscattered signal from the tag, the environment, and all other tags which may be in read range is also amplified, filtered and downconverted into complex baseband. The baseband signals are low-pass filtered and sampled with an oscilloscope, as is the original transmitted localization signal. Further processing is realized in MATLAB. Exemplary signals are shown in Section 6.

3.1. Derivation of the localization principle

Based on the result of the last equation (24) in Section 3, it is necessary to evaluate the channel response h_i regarding the distance between tag and reader. As stated at the beginning of Section 3 the localization should be performed using direct distance estimation, thus, the signal's time of flight t has to be evaluated in order to determine the distance d with the help of the propagation speed c, i.e.,

$$d = \frac{1}{2} c \cdot t, \tag{25}$$

with c usually being the propagation speed of electromagnetic waves complying to the speed of light. The factor one-half is introduced to compensate for the double distance the signal has to travel, i.e., from the reader to the tag and back.

In order to have a high positioning accuracy the signal must be broad regarding bandwidth, but the free spectrum for RFID, especially in Europe, is too small for that application. Therefore, higher out-of-band frequencies must be used. However, due to legal regulations, high bandwidth signals must be very low power, if applied. Ultra-wideband (UWB) signals [45] are such kind of signals and regulated by the Federal Communications Commission (FCC) and its counterparts in other regions, in order not to disturb any other in-band applications. UWB signals are defined as signals with bandwidths greater than 500 MHz or 20 % of the arithmetic mean of lower and upper cutoff frequency. The bandwidth used for the proposed system is 100 MHz due to the capability of UHF RFID tags working worldwide from around 840 MHz up to 960 MHz. Based on these conditions, although the proposed system only occupies 11 % of the arithmetic mean of the cutoff frequencies, the idea is to use low-power spreading signals for the ranging process. These signals are used to calculate the channel to a specific tag and back, thus extracting the time of flight information. As the low-power signals are hard to evaluate directly, the SNR is increased by performing coherent integration [46].

3.2. Mathematical model

Drawing up on Section 3, Equation (24), one can see that it is possible to derive the channel's impulse response upon evaluating the difference between both modulation states of the RFID transponder. Necessary for calculating the distance between tag and reader is the signal propagation time t of the up- and downlink channel. Multiplying half of t with the propagation speed results into the distance d between tag and reader (Equation (25)). As the bandwidth is limited to 100 MHz (Subsection 4.1), the pulse width is 10 ns minimum. Accordingly, a pulse width of 10 ns corresponds to a distance of around 3 m, supposing the speed of light in air is approximately 30 cm per nanosecond. Furthermore, the distance to be covered by this passive localization system is limited to the distance passive RFID tags are able to handle, which is, currently, limited to around 8 m [47]. In addition, the transmitted signal consists of more than one single pulse. These conditions lead to the fact, that the transmit signal and the receiving signal cannot be separated in time, as in ordinary RADAR applications. Another alternative is the principle of correlation, that can be used to determine the time shifted replica of the transmit pulse signal within the receiving signal [48]. The discrete correlation $R_{xy}[\tau]$ between two signals $x[t]$ and $y[t]$ is given with

$$R_{xy}[\tau] = \sum_{k=-\infty}^{+\infty} x[k] \cdot y[\tau + k] = x[t] \cdot y[t]. \tag{26}$$

The correlation term shows the time-shifted replicas of the signal $x[t]$ within the signal $y[t]$. A local maximum within the correlation term means a high correlation between $x[t]$ and $[t]$, i.e., a high linear match. The point in time of the maximum shows the time shift between $x[t]$ and $y[t]$, that is used to calculate the time between transmitted signal $x[t]$ and received signal $y[t]$.

3.3. Example

Let us derive the principle at a simplified example. Assuming the channel of the ith transponder is noise-free and multipath-free given with just

$$h_i[t] = a \cdot \delta[t - T_{\text{delay}}] \cdot e^{j\varphi_0}, \tag{27}$$

with a representing the reciprocal of the attenuation, $\delta[t - T_{\text{delay}}]$ the time delay T_{delay} of the channel with the Dirac delta function $\delta[t]$, and an initial phase shift of φ_0. Furthermore, the transponder modulation states $s_{i,1}$ and $s_{i,2}$ are given with 0 (full tag absorption) and 1 (full tag reflection). Equation (24) may now be written as

$$\Delta y_{\text{reader}}[t] = [s_{i,2} - s_{i,1}] * h_i[t] * x_{\text{reader}}[t] = [1 - 0] * a \cdot \delta[t - T_{\text{delay}}] \cdot e^{\varphi_0} * x_{\text{reader}}[t] = $$
$$= a \cdot \delta[t - T_{\text{delay}}] \cdot e^{j\varphi_0} * x_{\text{reader}}[t] \tag{28}$$

Performing the correlation to Equation (28) leads to

$$x_{\text{reader}}[t] \cdot \Delta y_{\text{reader}}[t] = x_{\text{reader}}[t] \cdot \left(a \cdot \delta[t - T_{\text{delay}}] \cdot e^{j\varphi_0} * x_{\text{reader}}[t] \right) \tag{29}$$

The convolution of $x_{\text{reader}}[t]$ with the time-shifted Dirac impulse $\delta[t - T_{\text{delay}}]$ delivers a time-shifted signal:

$$x_{\text{reader}}[t - T_{\text{delay}}] = \delta[t - T_{\text{delay}}] * x_{\text{reader}}[t] \tag{30}$$

The correlation of $x_{\text{reader}}[t - T_{\text{delay}}]$ with the original reader signal $x_{\text{reader}}[t]$ results in a time-shifted cross-correlation signal $R_{x_{\text{reader}}}[\tau - T_{\text{delay}}]$:

$$x_{\text{reader}}[t] \cdot \Delta y_{\text{reader}}[t] = a \cdot e^{j\varphi_0} \cdot R_{x_{\text{reader}}}[\tau - T_{\text{delay}}] \tag{31}$$

Performing the square of the absolute value to Equation (31), finally, leads to

$$|x_{\text{reader}}[t] \cdot \Delta y_{\text{reader}}[t]|^2 = |a|^2 \cdot |e^{j\varphi_0}|^2 \cdot |R_{x_{\text{reader}}}[\tau - T_{\text{delay}}]|^2 = |a|^2 \cdot |R_{x_{\text{reader}}}[\tau - T_{\text{delay}}]|^2 \tag{32}$$

The wanted time delay T_{delay} is evaluated by searching for the the maximum within the term $|x_{\text{reader}}[t] \cdot \Delta y_{\text{reader}}[t]|^2$:

$$T_{\text{delay}} = \underset{\tau}{\text{argmax}} \left\{ |a|^2 \cdot |R_{x_{\text{reader}}}[t - T_{\text{delay}}]|^2 \right\} \tag{33}$$

By receiving T_{delay} the distance d between reader and tag can be calculated by evaluating Equation (25) with

$$d = \frac{1}{2} \cdot c \cdot T_{\text{delay}}. \tag{34}$$

Multipath fading and receiver noise corrupt and distort the distance estimation. Gaussian noise on the low-power signals can be suppressed through coherent integration at the receiver. However, the increase in SNR due to integration is at the cost of receiving time [46]. The effect of multipath fading is more severe as it distorts the measurements in a way that is non-predictable without any a priori knowledge of the channel, which is given for a localization system working without channel prediction. The deployment of high-gain (low beam width) antennas with an electronic beam former can reduce the amount of multipath fading to an acceptable level.

4. Challenges and limitations

This section reveals the limitations and challenges of the proposed UHF RFID positioning system. Theoretical calculations show an accuracy limit at around 1 cm with the given hardware and signal limitations.

4.1. Limitations

The limitation of the system regarding the accuracy can be estimated using the Cramér-Rao Lower Bound (CRLB) [49], which defines a lower bound for an unbiased estimator $\hat{\theta}$. This means that the unbiased estimator of θ is always worse or equal to the CRLB. For an unbiased estimator $\hat{\theta}$ the standard deviation $\sigma_{\hat{\theta}}(\theta)$ is defined as [50]:

$$\sigma_{\hat{\theta}}(\theta) \geq \sqrt{\text{CRLB}_{\hat{\theta}}(\theta)} \tag{35}$$

Estimating the time-of-flight corresponds to the following CRBL definition of the standard deviation σ_x of the localization, i.e., the precision [50, 51]:

$$\sigma_x \geq \frac{c}{2\pi \cdot B_{RMS} \cdot \sqrt{2 \cdot \text{SNR}}} \tag{36}$$

$c$describes the propagation speed, SNR is the signal-to-noise ratio and B_{RMS} is the effective bandwidth of the signal used and is defined as

$$B_{RMS} = \sqrt{\int_B f^2 |S(f)|^2 df \Big/ \int_B |S(f)|^2 df} \tag{37}$$

with the Fourier transform of the signal $S(f)$ over the signal bandwidth B.

As the CRLB states in Equation (36), possible increases in precision are possible by either increasing the effective bandwidth of the signals or increasing the signal-to-noise ratio. If the given bandwidth is fixed, only an increase in SNR results in a higher measurement precision. As stated earlier, the SNR is increased by performing coherent integration. For instance, integration over $n = 10,000$ signals, results in an SNR increase of factor 10,000, but only in a precision increase of $\sqrt{10,000} = 100$. Theoretically, it is possible to increase the SNR as high as wanted, but receiver restrictions and timeouts limit the SNR to a certain level. These restrictions, mainly due to phase and quantization noise, define the limitations or the lower bounds of the localization system to a certain precision.

The applied hardware setup delivers the following SNR values for the quantization and phase noise. Thus, the receiver has an quantization error leading to an SNR of

$$SNR_{quantization} \approx 50 \text{ dB} = 10^5, \tag{38}$$

and phase noise leads to an SNR of

$$SNR_{phase} \approx 34 \text{ dB} = 2{,}512. \tag{39}$$

The total SNR is defined as

$$\frac{1}{SNR} = \frac{1}{SNR_{quantization}} + \frac{1}{SNR_{phase}} + \frac{1}{SNR_{signal}}. \tag{40}$$

SNR_{signal} is the power of the signal to the Gaussian noise power at the receiver. Figure 7 shows the maximum precision σ_x over certain SNR_{signal} values and coherent integrations with a factor of n. The effective bandwidth of the signal is given with $B_{RMS} = 36.66$ MHz. The SNR values and the effective bandwidth are derived from the receiver properties and the shape of the transmit pulse. Also, the factor one-half is considered due to half of the distance from tag to reader that reduces the precision σ_x in Equation (36) by a factor of 0.5.

As from Figure 7, it is shown that the lower bound for the standard deviation is around 1 cm. By increasing the number of coherent integrations n, the bound can be shifted to the left, which means, that the lower limit of the precision is reached for a lower SNR_{signal} value. For the proposed system, one measurement takes 1 μs, which increases to 1 s, if the coherent integration factor is $n = 1{,}000{,}000$.

4.2. Challenges

Challenges this localization system is facing are mainly:

- Multipath fading

- Non-constant tag reflection factors which vary by frequency and power

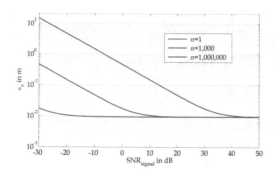

Figure 7. Cramér-Rao Lower Bound of the localization system

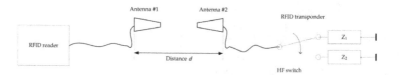

Figure 8. Measurement setup of RFID localization system

Multipath fading due to reflections, scattering and diffraction can be suppressed by using high-gain antennas with a high-focussed beam. Hence, electronic beam steering is necessary to cover the area to detect RFID tags. Using an omni-directional antenna avoids electronic beam steering at the cost of more multipath fading. Another alternative, to minimize multipath fading is the use of a much higher bandwidth. In future, UWB technology combined with RFID could have a major effect on improvements in positioning accuracy [52, 53].

The non-constant tag reflection factors that vary over frequency and power are able to strongly deteriorate the position estimation [16], if disregarded. One solution for this problem is revealed in [54].

5. Measurement results

This section shows the obtained measurement results. The first measurements are taken in an anechoic chamber, the second measurements are taken in an office environment. Both measurements are one-dimensional measurements.

5.1. Measurement setup

The measurement setup is given as in Figure 8. It consists of the reader unit as described in Section 3, an reader antenna (Antenna #1) and an RFID tag with tag antenna (Antenna #2) followed by a HF switch for emulating the tag modulation states with impedances Z_1 and Z_2. For the sake of simplicity Z_1 and Z_2 are chosen as *Short* and *Open*, i.e., $Z_1 = 0\ \Omega$ and $Z_2 = \infty\ \Omega$.

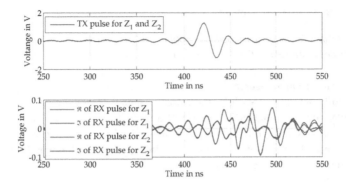

Figure 9. Exemplary transmit (TX) and receive (RX) pulses of the reader for Z_1 and Z_2, divided in real (R) and imaginary (I) signal components for an antenna to antenna distance of $d = 100$ cm

The measurement procedure is as follows. The HF switch toggles to impedance Z_1. Subsequently, the reader transmits and receives its signals as shown in Figure 9. Then, the switch toggles to Z_2 and, again, the reader transmits and receives its signals. Dependent on the number of coherent integrations, this procedure is repeated up to n. Finally, the sampled signals are evaluated in MATLAB. Figure 9 displays the transmit and receive signals for a given setup (anechoic chamber at a distance of 100 cm). The upper half of the figure shows the transmit signal – based on the Barker code [+1,-1] – used for both modulation states, Z_1 and Z_2. The lower half of the figure indicates the received signals for Z_1 and Z_2, respectively. As the received signals are complex-valued, real and imaginary parts are depicted for each RX signal. As seen in Figure 9, the received signals match each other for a certain period of time, until the difference in reflection (of Z_1 and Z_2) emerges (beginning at around 500 ns). These signals are used to determine the time shift between TX and RX signal and thus the distance between reader and tag. Evaluations of the correlations can be found in [48, 55].

The following two subsections show the measurement results, i.e., the result of the correlation difference, for two environments. First, a measurement in an anechoic chamber (Figure 10, left), second, a measurement in an office environment (Figure 10, right).

Figure 10. Measurement environments; left: anechoic chamber, right: office

5.2. Anechoic chamber measurements

The results of the measurement carried out in an anechoic chamber are depicted in Figure 11. The x-axis describes the real distances between the antennas, the y-axis describes the estimated distances. For normalization (cables, amplifiers, etc.) issues, the system is range-normalized to a distance of $d = 90$ cm (measurement with lowest variance). The coherent integration factor was chosen to be $n = 100$, i.e., each location was measured once with 100 transmit signals coherently integrated. The total RMSE error is 1.74 cm, which is the accuracy for a measurement distance from 80 cm to 280 cm. The fitting line in Figure 11 describes the regression line of the estimated distances. Hence, we can state that the system performs in the expected error ranges under very low multipath conditions.

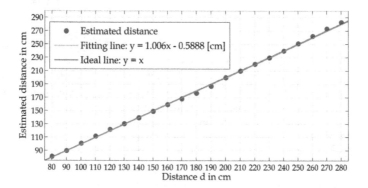

Figure 11. Results of the measurement carried out in anechoic chamber

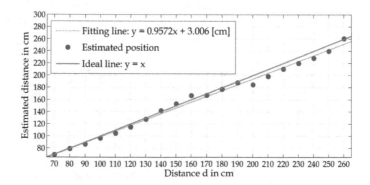

Figure 12. Results of the measurement carried out in an office environment

5.3. Office measurements

The results of the measurement carried out in the office environment are shown in Figure 12. Again, the x-axis describes the real distances between the antennas, the y-axis describes the estimated distances. The system is normalized to the distance of $d = 90$ cm, performed in the anechoic chamber. The coherent integration factor was chosen to be $n = 100$. The total RMSE error is 6.82 cm, which is the accuracy for a measurement distance from 70 cm to 260 cm. The fitting line describes the regression line of the estimated distances. The estimated values describe a nearly linear relation from 70 cm to 190 cm. The following estimated values are around 10 cm below the ideal line, the estimated value at the distance of 260 cm is back on track regarding the ideal line.

6. Result and discussion

The above measurements show that it is basically possible to gain range information down to accuracies of a few centimeters from the different modulation states of UHF RFID tags using wideband signals. However, there exist some simplifications, including the high-gain antennas and the tag modulation impedances given with open and short circuit (see also Subsection 5.2).

The idea behind the introduced localization system is based on the fact that current RFID-based localization systems either need a high effort in pre-calibration phases, suffer from bandwidth limitations, particularly in small frequency bands, e.g., as in Europe or need more complex hardware structures (phased array antennas) that only may be used in stationary, immobile applications. Therefore, a passive RFID-based positioning system should have ideally (Section 3) no change in hardware, high bandwidth, no pre-calibration phases and should be used in mobile applications. The suggested system includes these issues in the following way. There is no pre-calibration phase necessary as the system uses direct range estimation. This, however, is only possible due to the high bandwidth used along with low-power signals to stay within the required power spectrum densities. Changes in hardware would incorporate high bandwidth filter structures, a fast signal generator for the transmit pulses and a high accurate A/D converter for the incoming signals. Finally, it can be stated that such a localization system for mobile indoor positioning is possible, if the required hardware prerequisites are created.

7. Summary and conclusion

This chapter dealt with the concepts of localization comprising primarily UHF and microwave RFID systems. After describing the fundamental principles behind localization, a survey was given for state-of-the-art RFID localization systems. Subsequently, a novel RFID localization system using wideband signals was introduced. A theoretical derivation of the range determination was given in Section 4, whereas Section 5 revealed the limits and challenges of the proposed localization system, e.g., through evaluation of the Cramér-Rao lower bound.

Finally, measurement results carried out in different environments (anechoic chamber, office) showed that the proposed system works within the former deduced limitations. The measurements showed a one-dimensional accuracy (RMSE) of 1.7 cm in the anechoic chamber, and an accuracy (RMSE) of 6.8 cm within the office environment. Tag reflection normalization and the usage of omni-directional antennas along with real-time localization are subjects to future work.

Acknowledgements

The authors would like to thank their colleagues from the Chair of Information Technology as well as from the Fraunhofer Institute for Integrated Circuits. Special thanks to our colleagues Frederik Beer, Gerd Kilian and Hendrik Lieske from the telemetry group for their valuable feedback.

Author details

Andreas Loeffler[1*] and Heinz Gerhaeuser[2]

*Address all correspondence to: loeffler@like.eei.uni-erlangen.de

1 Chair of Information Technology (Communication Electronics) Engineering, Friedrich-Alexander-University of Erlangen-Nuremberg, Erlangen, Germany

2 Fraunhofer Institute for Integrated Circuits IIS, Erlangen, Germany

References

[1] Kaplan, E. D, & Hegarty, C. J. Understanding GPS: Principles And Applications. Artech House Mobile Communications Series. Artech House; (2006).

[2] Retscher, G, & Kealy, A. . Ubiquitous Positioning Technologies for Modern Intelligent Navigation Systems. The Journal of Navigation. (2006). ; . Available from: http://dx.doi.org/10.1017/S0373463305003504., 59(01), 91-103.

[3] Chen, Z, Xia, F, Huang, T, Bu, F, & Wang, H. . A localization method for the Internet of Things. The Journal of Supercomputing. (2011). ; . 10.1007/s11227-011-0693-2. Available from: http://dx.doi.org/10.1007/s11227-011-0693-2., 1-18.

[4] Xiang, Z, Song, S, Chen, J, Wang, H, Huang, J, & Gao, X. A wireless LAN-based indoor positioning technology. IBM Journal of Research and Development. (2004). sep;48(5.6): 617-626.

[5] Mautz, R. . Overview of current indoor positioning systems. Geodezija ir Kartografija. (2009). ; . Available from: http://www.tandfonline.com/doi/abs/ 10.3846/1392-1541.2009.35.18-22., 35(1), 18-22.

[6] Ashton, K. . Whither the Five-Cent Tag? RFID Journal; (2011). . Available from: http:// www.rfidjournal.com/article/view/8212.

[7] Escribano, J. G, Garcia, A, Wissendheit, U, Loeffler, A, & Pastor, J. M. Analysis of the applicability of RFID & wireless sensors to manufacturing and distribution lines trough a testing multi-platform. In: Industrial Technology (ICIT), 2010 IEEE International Conference on; (2010). , 1379-1385.

[8] Baars, H, Gille, D, & Strüker, J. Evaluation of RFID applications for logistics: a framework for identifying, forecasting and assessing benefits. European Journal of Information Systems. (2009). , 18(6), 578-591.

[9] Desmons, D. . UHF Gen2 for item-level tagging. Presentation at RFID World. (2006). ;Available from: http://www.impinj.com/files/Impinj_ILT_RFID_WORLD.pdf.

[10] Dobkin, D. M. The RF in RFID: Passive UHF RFID in Practice. Communications Engineering Series. Elsevier / Newnes; (2007).

[11] Wackerly, D. D, Mendenhall, W, & Scheaffer, R. L. Mathematical Statistics with Applications. Thomson, Brooks/Cole; (2008).

[12] Manolakis, D. E. Efficient solution and performance analysis of 3-D position estimation by trilateration. Aerospace and Electronic Systems, IEEE Transactions on. (1996). oct;, 32(4), 1239-1248.

[13] Murphy, W. S. Determination of a Position Using Approximate Distances and Trilateration. Colorado School of Mines; (2007).

[14] Navidi, W. , Jr WSM, Hereman W. Statistical methods in surveying by trilateration. Computational Statistics & Data Analysis. (1998). ; . Available from: http:// www.sciencedirect.com/science/article/pii/S0167947397000534., 27(2), 209-227.

[15] Hoene, C, & Willmann, J. Four-way TOA and software-based trilateration of IEEE 802.11 devices. In: Personal, Indoor and Mobile Radio Communications, 2008. PIMRC 2008. IEEE 19th International Symposium on; (2008). , 1-6.

[16] Arnitz, D, Muehlmann, U, Semiconductors, N, & Witrisal, K. Tag-Based Sensing and Positioning in Passive UHF RFID: Tag Reflection. In: 3rd Int. EURASIP workshop on RFID Technology; (2010).

[17] Liu, H, Darabi, H, Banerjee, P, & Liu, J. Survey of Wireless Indoor Positioning Techniques and Systems. Systems, Man, and Cybernetics, Part C: Applications and Reviews, IEEE Transactions on. (2007). nov;, 37(6), 1067-1080.

[18] Honkavirta, V, Perala, T, Ali-loytty, S, & Piche, R. A comparative survey of WLAN location fingerprinting methods. In: Positioning, Navigation and Communication, 2009. WPNC 2009. 6th Workshop on; (2009). , 243-251.

[19] Seitz, J, Vaupel, T, Jahn, J, Meyer, S, Boronat, JG, & Thielecke, J. . A Hidden Markov Model for urban navigation based on fingerprinting and pedestrian dead reckoning. In: Information Fusion (FUSION), 2010 13th Conference on; (2010). . . Available from: http://ieeexplore.ieee.org/stamp/stamp.jsp?tp=&arnumber=5712025., 1-8.

[20] Meyer, S. Feldstärkemessung; (2011).

[21] Nikitin, P. V, Martinez, R, Ramamurthy, S, Leland, H, & Spiess, G. Rao KVS. Phase based spatial identification of UHF RFID tags. In: RFID, 2010 IEEE International Conference on; (2010). , 102-109.

[22] Sanpechuda, T, & Kovavisaruch, L. A review of RFID localization: Applications and techniques. In: Electrical Engineering/Electronics, Computer, Telecommunications and Information Technology, 2008. ECTI-CON 2008. 5th International Conference on. (2008). , 2, 769-772.

[23] Zhang, Y, Li, X, & Amin, M. . In: Principles and Techniques of RFID Positioning. John Wiley & Sons, Ltd; (2010). . . Available from: http://dx.doi.org/ 10.1002/9780470665251.ch15., 389-415.

[24] Vossiek, M, Miesen, R, Wittwer, J, & Identification, R. F. and localization- recent steps towards the internet of things in metal production and processing. In: Microwave Radar and Wireless Communications (MIKON), 2010 18th International Conference on; (2010). , 1-8.

[25] Hightower, J, Want, R, & Borriello, G. SpotON: An indoor 3D location sensing technology based on RF signal strength. UW CSE 00-02-02, University of Washington, Department of Computer Science and Engineering, Seattle, WA. (2000). Available from: ftp://128.95.1.178/tr/2000/02/UW-CSE-00-02-02.pdf.

[26] Seidel, S. Y, & Rappaport, T. S. MHz path loss prediction models for indoor wireless communications in multifloored buildings. Antennas and Propagation, IEEE Transactions on. (1992). feb;, 40(2), 207-217.

[27] Bechteler, T. F, & Yenigun, H. D localization and identification based on SAW ID-tags at 2.5 GHz. Microwave Theory and Techniques, IEEE Transactions on. (2003). may;, 51(5), 1584-1590.

[28] Stelzer, A, Pourvoyeur, K, & Fischer, A. Concept and application of LPM- a novel 3-D local position measurement system. Microwave Theory and Techniques, IEEE Transactions on. (2004). dec;, 52(12), 2664-2669.

[29] Stove, A. G. Linear FMCW radar techniques. Radar and Signal Processing, IEE Proceedings F. (1992). oct;, 139(5), 343-350.

[30] Vossiek, M, Roskosch, R, & Heide, P. Precise 3-D Object Position Tracking using FMCW Radar. In: Microwave Conference, 1999. 29th European. (1999). , 1, 234-237.

[31] Heidrich, J, Brenk, D, Essel, J, Fischer, G, Weigel, R, & Schwarzer, S. Local positioning with passive UHF RFID transponders. In: Wireless Sensing, Local Positioning, and

RFID, 2009. IMWS 2009. IEEE MTT-S International Microwave Workshop on; (2009). , 1-4.

[32] Li, X, Zhang, Y, & Amin, M. G. Multifrequency-based range estimation of RFID Tags. In: RFID, 2009 IEEE International Conference on; (2009). , 147-154.

[33] Ni, L. M, Liu, Y, Lau, Y. C, & Patil, A. P. LANDMARC: indoor location sensing using active RFID. Wireless networks. (2004). , 10(6), 701-710.

[34] Hightower, J, Vakili, C, Borriello, G, & Want, R. Design and calibration of the spoton ad-hoc location sensing system. unpublished, August. (2001).

[35] Chattopadhyay, A, & Harish, A. R. Analysis of low range Indoor Location Tracking techniques using Passive UHF RFID tags. In: Radio and Wireless Symposium, 2008 IEEE. IEEE; (2008). , 351-354.

[36] Chattopadhyay, A, & Harish, A. R. Analysis of UHF passive RFID tag behavior and study of their applications in low range indoor location tracking. In: Antennas and Propagation Society International Symposium, 2007 IEEE; (2007). , 1217-1220.

[37] Park, S, & Lee, H. Self-recognition of Vehicle Position using UHF Passive RFID Tags. (2012).

[38] Hahnel, D, Burgard, W, Fox, D, Fishkin, K, & Philipose, M. Mapping and localization with RFID technology. In: Robotics and Automation, 2004. Proceedings. ICRA'04. 2004 IEEE International Conference on. IEEE; (2004). , 1, 1015-1020.

[39] Hahnel, D, Burgard, W, Fox, D, & Thrun, S. An efficient fastSLAM algorithm for generating maps of large-scale cyclic environments from raw laser range measurements. In: Intelligent Robots and Systems, 2003. (IROS 2003). Proceedings. 2003 IEEE/RSJ International Conference on. (2003). vol.1., 1, 206-211.

[40] Dellaert, F, Fox, D, Burgard, W, & Thrun, S. Monte Carlo localization for mobile robots. In: Robotics and Automation, 1999. Proceedings. 1999 IEEE International Conference on. (1999). vol.2., 2, 1322-1328.

[41] Parlak, S, & Marsic, I. Non-intrusive localization of passive RFID tagged objects in an indoor workplace. In: Proc. IEEE Int RFID-Technologies and Applications (RFID-TA) Conf; (2011). , 181-187.

[42] Friedman, J, Hastie, T, & Tibshirani, R. Additive logistic regression: a statistical view of boosting (With discussion and a rejoinder by the authors). The annals of statistics. (2000). , 28(2), 337-407.

[43] Karmakar, N. C, Roy, S. M, & Ikram, M. S. Development of Smart Antenna for RFID Reader. In: RFID, 2008 IEEE International Conference on; (2008). , 65-73.

[44] Finkenzeller, K. RFID Handbook: Fundamentals and Applications in Contactless Smart Cards, Radio Frequency Identification and Near-Field Communication. Wiley; (2010).

[45] Taylor, J. D. Introduction to ultra-wideband radar systems. CRC; (1995).

[46] Loeffler, A, & Gerhaeuser, H. A Novel Approach for UHF-RFID-Based Positioning Through Spread- Spectrum Techniques. Smart Objects: Systems, Technologies and Applications (RFID Sys Tech), 2010 European Workshop on. (2010). june;, 1-10.

[47] Ussmueller, T, Brenk, D, Essel, J, Heidrich, J, Fischer, G, & Weigel, R. A multistandard HF/ UHF-RFID-tag with integrated sensor interface and localization capability. In: RFID (RFID), 2012 IEEE International Conference on; (2012). , 66-73.

[48] Loeffler, A. Localizing passive UHF RFID tags with wideband signals. In: Microwaves, Communications, Antennas and Electronics Systems (COMCAS), 2011 IEEE International Conference on; (2011). , 1-6.

[49] Gustafsson, F, & Gunnarsson, F. Mobile positioning using wireless networks: possibilities and fundamental limitations based on available wireless network measurements. Signal Processing Magazine, IEEE. (2005). july;, 22(4), 41-53.

[50] Fowler, M. . EECE 522 Estimation Theory; (2012). . Available from: http://www.ws.binghamton.edu/fowler/fowler%20personal%20page/EE522.htm.

[51] Gezici, S. . A Survey on Wireless Position Estimation. Wireless Personal Communications. (2008). ; . 10.1007/s11277-007-9375-z. Available from: http://dx.doi.org/10.1007/s11277-007-9375-z., 44, 263-282.

[52] Arnitz, D, Adamiuk, G, Muehlmann, U, & Witrisal, K. . UWB channel sounding for ranging and positioning in passive UHF RFID. 11th COST2100 MCM. (2010). ;Available from: http://spsc.tu-graz.ac.at/system/files/arnitzcostmcm10.pdf.

[53] Arnitz, D, Muehlmann, U, & Witrisal, K. UWB ranging in passive UHF RFID: proof of concept. Electronics Letters. (2010). , 46(20), 1401-1402.

[54] Viikari, V, Pursula, P, & Jaakkola, K. Ranging of UHF RFID Tag Using Stepped Frequency Read-Out. Sensors Journal, IEEE. (2010). sept;, 10(9), 1535-1539.

[55] Loeffler, A. Dispersion Effects at High Bandwidth Localization for UHF RFID Tags. ITG-Fachbericht-Smart SysTech 2012. (2012).

Chapter 6

Design of a Zeroth Order Resonator UHF RFID Passive Tag Antenna with Capacitive Loaded Coplanar Waveguide Structures

Muhammad Mubeen Masud and
Benjamin D. Braaten

Additional information is available at the end of the chapter

1. Introduction

The use and development of Radio Frequency Identification (RFID) systems has undergone substantial growth in the past decade in many new areas. Some of these areas include wireless sensor systems, metamaterials and compact antennas [1-8]. However, much of this new growth has required more performance from traditional passive RFID systems. In particular, the need for more compact antennas with performances comparable to much larger resonant antennas is one such condition. To fulfill the requirements of compact antennas, researchers have developed various novel RFID antenna designs [2-4], including metamaterial-based RFID antenna designs [1,5-8] to improve the performance of RFID systems. Using composite right/left-handed (CRLH) transmission line (TL) based metamaterials to show the unique property of zeroth-order resonance (ZOR) [9,10] is one such method to reduce the overall size of an antenna. More specifically, a ZOR-TL can be used to make an electrically small antenna to appear electrically large; which leads to improved matching and radiation properties. This is done by producing a zero phase constant at a non-zero frequency (i.e. the wavelength of the travelling wave becomes infinite) on the TL. This is a unique property which makes the resonance condition independent from the physical dimensions of the antenna or TL [11-13] so it can be used to design miniature antennas for passive UHF RFID applications. The resonance of such antennas at any operating frequency only depend on its CRLH characteristics to acquire ZOR at that frequency and less to do with the physical dimensions of corresponding antenna.

This chapter will focus on the design of ZOR antennas for passive UHF RFID tags. First, a brief introduction and working principles of RFID systems is presented using Friis's transmission equation. Then, the characteristics of CRLH transmission lines will be discussed and its Bloch impedance will be derived to introduce the ZOR concept. Then coplanar-waveguides (CPW) and its characteristics are presented. Then the design of a capacitive loaded CPW based ZOR antenna for passive UHF RFID tag is discussed. Finally, future work and conclusion about this chapter is presented.

2. Introduction to RFID systems

RFID technology has drawn great attention in the past decade. Recently it has been used in inventory control, managing large volumes of books in libraries and tracking of products in the retail supply chain [14,15]. Its usage is growing and replacing the bar code technology used for the purpose of object identification and recognition. A bar code requires a clear line of sight and a small distance between the object and the laser bar code scanner (which is a limitation) whereas RFID works at microwave frequencies so it can identify the object from a distance, it does not require line of sight for its operation and unlike bar codes it can also store some additional information which makes it very attractive as compared to bar codes [1].

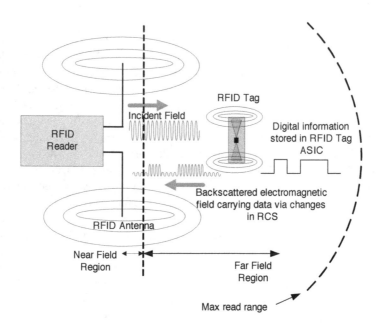

Figure 1. Overview of RFID System

A RFID system consists of a RFID reader and a RFID tag. An overview of a typical RFID system is shown in Fig. 1. RFID systems comprise of RFID tags or transponders which are fairly simple, small and inexpensive devices at one end and a reader which is relatively complex and a bigger device on the other end. Application Specific Integrated Circuits (ASICs) are attached to the tag antenna and are used for sensor applications, to harvest energy, communicate and store information for later recovery. The reader emits an electromagnetic field which contains power and timing information for use by the passive RFID. If a RFID tag comes within the range (also known as the interrogation zone [1]) it receives the information which is fed to the ASIC and in response the ASIC switches its impedance states between a lower and higher value in a predetermined fashion as shown in Fig. 2. By changing the impedance states the ASIC changes the radar cross-section (RCS) of the tag antenna thus changing the backscattered power. This backscattered power is collected at the reader and is used for tag identification and information. The maximum distance for which a reader can successfully identify a tag is known as max read range.

RFID tags are usually classified into three categories: active tags, semi-passive tags and passive tags [1]. An active tag has a dedicated power supply for operation on the tag. A semi-passive tag has an integrated power supply attached to it and it only starts working when electromagnetic power transmitted by the reader is incident on the tag. This feature enhances the maximum read range of the tag [1] because less power is required from the incoming incident field from the reader. A passive tag has no power source attached to it and it harvests power for its operation from the incident electromagnetic field transmitted by the reader.

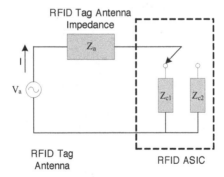

Figure 2. Thevenin equivalent circuit of RFID tag

A common method to describe the RFID wireless communication system is the following Friis transmission equation [16]:

$$P_r = P_t \frac{G_r G_t \lambda^2}{(4\pi R)^2} q \tag{1}$$

where P_r is the power received by RFID tag, P_t is power transmitted by RFID reader, G_r is the gain of tag antenna, G_t is the gain of reader, λ is free space wavelength of the operating frequency of reader, R is distance between reader and tag and q is impedance mismatch factor ($0 \le q \le 1$) between impedance of the antenna on the tag and the input impedance of the ASIC on the tag. Equation (1) assumes a perfect polarization match between the antenna on the reader and the antenna on the RFID tag. Reorganizing (1) and solving for R, the following equation for determining the read rang of a tag can be derived [17,18] as:

$$R = \frac{\lambda}{4\pi}\sqrt{\frac{qG_tG_rP_t}{P_r}} \tag{2}$$

If the minimum power required for tag operation is P_{th} then Equation (2) can be written as

$$R_{max} = \frac{\lambda}{4\pi}\sqrt{\frac{qG_tG_rP_t}{P_{th}}} \tag{3}$$

Equation (3) is useful for designers to determine the maximum operating range of the tag. Typically the approach by a designer is to maximize the R_{max}. One way of achieving this is to minimize the mismatch between tag antenna and ASIC impedances or design a receive antenna on the RFID tag with a maximized gain G_r.

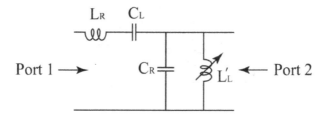

Figure 3. Reconfigurable CRLH-TL

3. Introduction to left-handed propagation

To help illustrate the use of ZOR properties to improve the gain and matching of a compact antenna on a passive UHF RFID tag, several properties of left-handed (LH) propagation will be introduced and summarized here. It is well known that the equivalent circuit of a traditional printed microstrip TL consists of a series inductance and a shunt capacitance. The series inductance is caused by the current travelling down the printed TL and the shunt capacitance represents the capacitance between the printed signal conductors on one side of the board and the reference or ground plane. In fact, this inductance and capacitance exists on every printed TL (traditional or CRLH) because in the propagating band current is trav-elling down the TL and there is always capacitance between the conductors supporting this current and a reference conductor. When introducing the CRLH-TL, this series inductance and shunt capacitance is referred to as the parasitic values and are denoted in Fig. 3 as L_R and C_R. The subscript R stands for right-handed (RH) propagation.

Next, to support LH-propagation, a series capacitance and a shunt inductance is introduced. These values are shown in Fig. 3 and are denoted C_L and L_L, respectively. The subscript L stands for left-handed propagation. More particularly, the series capacitance is in series with the inductance and the shunt inductance is in parallel with the shunt capacitance. Therefore, to achieve LH-propagation, C_L and L_L should dominate over the values of L_R and C_R. Closer observation of the equivalent circuit in Fig. 3 shows that the LH-values will only dominate over a certain band which is called the LH-propagating band. When the RH-values of L_R and C_R are dominant, this is called the RH-propagating band. When both the RH- and LH-values are equal; this is called the transition frequency between the RH- and LH-propagat-ing bands or simply the transition frequency. In practice, the series capacitance is usually introduced by defining interdigital capacitors down the length of the TL [10]. The shunt in-ductance has been introduced in many different ways such as split ring resonators and shunt stubs [10].

A CRLH-TL has several unique properties as a result of the introduction of C_L and L_L. The property used in this work is the sign change associated with the phase constant. The phase constant on a CRLH-TL is opposite to the phase constant on conventional RH-TL. This phase advance feature can be very useful for antenna designers and will be used in the next few sections to introduce the idea of ZOR antennas.

4. Coplanar-waveguide structures

The term "Coplanar" means sharing the same plane and this is the type of transmission line where the reference conductors are in the same plane as of signal carrying conductor. The signal carrying conductor is placed in the middle with a reference plane conductor on either side as shown in Fig. 4. The advantage of having both conductors in the same plane lies in the fact that it is easier to mount lumped components between the two planes and it is easier to realize shunt and series configurations. The CPW was first proposed by Wen [19] and since then have been used extensively in wireless communications [20,21].

The disadvantage of CPW is that it can be difficult to maintain the same potential between the reference and signal conductors throughout the signal trace. Nevertheless many advan-ces have been made by using CPW such as novel filters [22] and right/left handed propaga-tion on CPW lines [23].

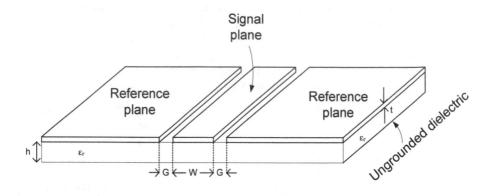

Figure 4. CPW transmission line on ungrounded dielectric

Several properties of the CPW-TL in Fig. 4 are derived next. These expression will be used later to describe the ZOR-RFID antenna. The attenuation and phase constants can be derived by performing a quasi-static analysis of a CPW [24]. The phase velocity and characteristic impedance equations can be written as [24]:

$$v_{cp} = \left(\frac{2}{\varepsilon_r + 1}\right)^{1/2} c \tag{4}$$

and

$$Z_{0cp} = \frac{30\pi}{\sqrt{\varepsilon_{re}^t}} \frac{K(k_e')}{K(k_e)} \tag{5}$$

where

$$k_e = \frac{W_e}{(W_e + 2G_e)} \cong k + \frac{(1 - k^2)\Delta}{2G} \tag{6}$$

$$k = \frac{W}{W + 2G} \tag{7}$$

$$\Delta = (1.25t / \pi)[1 + \ln(4\pi W / t)] \tag{8}$$

$$k' = (1 - k^2)^{1/2} \tag{9}$$

$$\varepsilon_{re}^{\,t} = \varepsilon_{re} - \frac{0.7(\varepsilon_{re} - 1)t\,/\,G}{\left[\frac{K(k)}{K(k')}\right] + \frac{0.7t}{G}}$$ (10)

and

$$\varepsilon_{re} = \frac{\varepsilon_r + 1}{2}\left[tanh\,\{1.785\log\,(h\,/\,G) + 1.75\} + \frac{kG}{h}\{0.04 - 0.7k + 0.01(1 - 0.1\varepsilon_r)(0.25 + k)\}\right]$$ (11)

Here W is the width of the center conductor, G is the spacing between the center conductor and the reference conductor, ε_r is the relative permittivity of the dielectric, c is the speed of light and t is the thickness of the conductor. K(k) is the complete elliptic integral of the first kind and the ratio K(k)/K(k') has been reported in [24,25] as:

$$\frac{K(k')}{K(k)} = \frac{1}{\pi}ln\left[2\frac{1 + \sqrt{k}}{1 - \sqrt{k}}\right] for\,\,0.707 \le k \le 1$$ (12)

and

$$\frac{K(k')}{K(k)} = \frac{\pi}{ln\left[2\frac{1 + \sqrt{k'}}{1 - \sqrt{k'}}\right]}\,for\,\,0 \le k \le 0.707$$ (13)

Using equations (4)-(13) the attenuation constant due to ohmic losses can be calculated as [24]:

$$\alpha_c^{cw} = 4.88*10^{-4}R_s\varepsilon_{re}Z_{0cp}\frac{P'}{\pi G}\left(1 + \frac{W}{G}\right)\left\{\frac{\frac{1.25}{\pi}ln\frac{4\pi W}{t} + 1 + \frac{1.25t}{\pi W}}{\left[2 + \frac{W}{G} - \frac{1.25t}{\pi G}\left(1 + ln\frac{4\pi W}{t}\right)\right]^2}\right\}dB\,\Big|\,unit\,length$$ (14)

where

$$P' = \left(\frac{K}{K'}\right)^2 P$$ (15)

$$P = \begin{cases} \dfrac{k}{\left(1 - \sqrt{1 - k^2}\right)(1 - k^2)^{3/4}}\,for\,\,0.0 \le k \le 0.707 \\[3mm] \dfrac{1}{(1 - k)\sqrt{k}}\left(\dfrac{K}{K'}\right)^2\,for\,\,0.707 \le k \le 1.0 \end{cases}$$ (16)

and

$$R_s = \sqrt{\rho\pi f\mu}$$ (17)

The attenuation constant due to dielectric losses is [24]:

$$\alpha_d = 27.3 \frac{\varepsilon_r}{\sqrt{\varepsilon_{re}}} \frac{\varepsilon_{re} - 1}{\varepsilon_r + 1} \frac{\tan \delta}{\lambda_0} \; dB \Big/ \; unit \; length \tag{18}$$

Here tan(δ) is the loss tangent of the dielectric and the total attenuation can be written as:

$$\alpha_{cwp} = \alpha_c + \alpha_d \tag{19}$$

Thus, the phase constant can be calculated as [20]:

$$\beta_{cpw} = \frac{2\pi f}{v_{cp}} \tag{20}$$

Next, these expressions will be used to introduce the interdigital capacitor loaded CPW which will then be used to design a ZOR-RFID antenna.

5. Interdigital capacitor loaded CPW

An Interdigital capacitor loaded transmission line provides a series resonance. The Zeroth Order Resonance (ZOR) of an interdigital capacitor loaded CPW has been investigated and reported in [26]. The equivalent transmission line model of an interdigital capacitor loaded transmission line is shown in Fig. 5 and consists of two symmetric transmission lines interconnected with a series capacitance. The host transmission line has been shown equally divided into two parts. Since the size of the unit cell is much smaller than the guided wavelength, the transmission line can be modeled with an equivalent circuit with a series inductance and shunt capacitance (as discussed in Section 3).

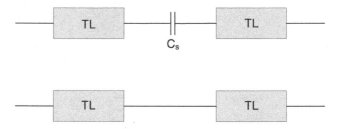

Figure 5. Equivalent circuit model of interdigital capacitor loaded

The geometry (layout) of the interdigital capacitor based unit cell is shown in Fig. 6. The capacitance between the interdigital capacitor and bilateral ground plane is fairly small as compared to the series capacitance of the interdigital capacitor so it can be neglected. This unit cell can be repeated periodically to design the ZOR antenna.

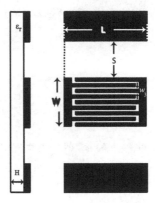

Figure 6. Interdigital capacitor loaded CPW unit cell

Since the unit cell will be repeated periodically and will be symmetric about the port of the antenna, it will resemble the TL in Fig. 5. Therefore, the propagation constant γ (where $\gamma = \alpha + j\beta$) and characteristic impedance (also known as block impedance) Z_B can be expressed in terms of an ABCD matrix as [20]:

$$\cosh \gamma L = A \tag{21}$$

and

$$Z_B = \frac{B Z_0}{\sqrt{A^2 - 1}} \tag{22}$$

Here L is the length of the unit cell and Z_0 is the characteristic impedance of the CPW. The propagation constant of the TL is $\gamma_{CPW} = \alpha_{CPW} + \beta_{CPW}$ where α_{CPW} and β_{CPW} can be calculated from (19) and (20), respectively.

Next, the ABCD matrix of the circuit shown in Fig. 5 can be determined as [20]:

$$\begin{bmatrix} A & B \\ C & D \end{bmatrix}_{CPW} = \begin{bmatrix} \cosh \dfrac{\gamma_{CPW} L}{2} & Z_0 \sinh \dfrac{\gamma_{CPW} L}{2} \\ Y_0 \sinh \dfrac{\gamma_{CPW} L}{2} & \cosh \dfrac{\gamma_{CPW} L}{2} \end{bmatrix} \tag{23}$$

and

$$\begin{bmatrix} A & B \\ C & D \end{bmatrix}_{inter\text{-}digital\ capacitor} = \begin{bmatrix} 1 & \dfrac{1}{j\omega C} \\ 0 & 1 \end{bmatrix} \tag{24}$$

Here L/2 represents half of the CPW length. The ABCD matrix of the whole unit cell can be calculated from (23) and (24) as:

$$
\begin{bmatrix} A & B \\ C & D \end{bmatrix} = \begin{bmatrix} A & B \\ C & D \end{bmatrix}_{CPW} * \begin{bmatrix} A & B \\ C & D \end{bmatrix}_{inter\text{-}digital\ capacitor} * \begin{bmatrix} A & B \\ C & D \end{bmatrix}_{CPW} \tag{25}
$$

From (25), parameter A can be calculated and (21) can be written as:

$$
\cosh \alpha L \, \cos \beta L \; + \; j\sinh \alpha L \, \sin \beta L \; = M + jN + \frac{1}{j2Z_0\omega C}(O + jP) \tag{26}
$$

where

$$
M = \cosh \alpha_{CPW} L \, \cos \beta_{CPW} L \tag{27}
$$

$$
N = \sinh \alpha_{CPW} L \, \sin \beta_{CPW} L \tag{28}
$$

$$
O = \sinh \alpha_{CPW} L \, \cos \beta_{CPW} L \tag{29}
$$

and

$$
P = \cosh \alpha_{CPW} L \, \sin \beta_{CPW} L \tag{30}
$$

In (26) α represents the attenuation constant and β represents the phase constant of the Bloch wave propagating on the unit cell whereas α_{CPW} and β_{CPW} are attenuation and phase constants, respectively, of the host CPW. From (26) the real and imaginary parts can be separated which gives:

$$
\cosh \alpha L \, \cos \beta L \; = \cosh \alpha_{CPW} L \, \cos \beta_{CPW} L \; + \frac{\cosh \alpha_{CPW} L \, \sin \beta_{CPW} L}{2Z_0\omega C} \tag{31}
$$

and

$$
\sinh \alpha L \, \sin \beta L \; = \sinh \alpha_{CPW} L \, \sin \beta_{CPW} L \; - \frac{\sinh \alpha_{CPW} L \, \cos \beta_{CPW} L}{2Z_0\omega C} \tag{32}
$$

The unknowns in (31) and (32) are α and β of the Bloch wave. Solving for α and β gives:

$$
\alpha = \frac{1}{L} \cosh^{-1}\left(\frac{\sqrt{Q^2 + (R+1)^2} + \sqrt{Q^2 + (R-1)^2}}{2} \right) \tag{33}
$$

and

$$\beta = \frac{1}{L} cos^{-1}\left(\frac{\sqrt{Q^2 + (R + 1)^2} - \sqrt{Q^2 + (R - 1)^2}}{2}\right)$$ (34)

where Q and R are the right hand sides of (31) and (32), respectively. The key idea when designing a ZOR antenna is to determine the frequency at which equation (34) is equal to zero. Since the propagation constant is inversely proportional to the wavelength, when equation (34) is zero, the wavelength at that frequency is equal to infinity. At this frequency, the antenna looks infinitely long electrically. In the next section, the expressions derived here for the interdigital capacitor loaded CPW will be used to design a ZOR-RFID antenna.

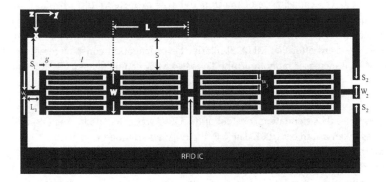

Figure 7. Layout of proposed ZOR RFID antenna with capacitor loaded CPW [30]

5.1. Zeroth order resonance

The layout of the proposed ZOR RFID antenna is shown in Fig. 7 [30]. The port of the antenna is located in the middle of the antenna with series capacitance down each arm. The operating principle of this antenna is based on the capacitive input impedance of the passive RFID ASIC. At resonance, the interdigital capacitors are supporting a wave propagating along the antenna. Since the input impedance of the ASIC is also capacitive the ASIC also supports wave propagation along the antenna in a manner similar to the interdigital capacitors [30]. During this process, the ASIC harvests the required power to perform the desired tasks and communicate while simultaneously supporting the wave propagating on the antenna.

The first step in the design process is to determine what capacitance is required to equate β to zero at the desired operating frequency such that the antenna looks infinitely long. For discussion, the non-zero frequency at which β becomes zero is known as the zeroth order resonance (ZOR) frequency [26], [30]. For simplicity a lossless ($\alpha = 0$) CPW line is assumed and then from (31) the required capacitance can be calculated to achieve ZOR at a particular design frequency as:

$$C = \frac{\cosh \alpha_{CPW} L \; \sin \beta_{CPW} L}{2\omega_r Z_0 (1 - \cosh \alpha_{CPW} L \; \cos \beta_{CPW} L \;)} \tag{35}$$

Since we are interested in designing a ZOR antenna for the passive UHF RFID band, 915 MHz is taken as the operating frequency and from (35) the required capacitance can be calculated as C = 2.64 pF.

The unit cell shown in Fig. 7 was simulated in ADS 2009 with design parameters L = 17.56 mm, W = 8.82 mm, w_3 = 0.36 mm, S = 7.96 mm and H = 1.524 mm. A Rogers TMM4 (ε_r = 4.5 and tan δ = 0.002) was used as a substrate. For the lossless case the attenuation constant of the CPW and loss tangent of the substrate was assumed to be zero and a perfect conductor was considered. The capacitance of the unit cell was extracted [10] to be $C_{extracted}$ = 2.4 pF which is close to the required capacitance for ZOR at 915 MHz. The dispersion characteristics are plotted in Fig. 8. It can be noted that the attenuation constant decreases monotonically and becomes zero after 944 MHz. Similarly the propagation constant remains zero and after 944 MHz it increases monotonically. Thus 944 MHz can be taken as ZOR frequency for the given unit cell which comes within 3.2% of the required resonance frequency of 915 MHz. More discussion on this is reported in [30].

For the lossy case the attenuation constant of the CPW was calculated using (19) and the loss tangent was taken as tan δ = 0.002. The conductivity was defined as σ = 5.8 x 10⁷ S/m with a conductor thickness of 35 μm. The dispersion characteristics for the lossy case were also presented in Fig. 8. A similar response for both the lossy and lossless case is shown except for the fact that the phase constant is non-zero below the ZOR point and similarly the attenuation constant is non-zero after the ZOR point. Here the ZOR point is taken as the point at which α = β and it coincides with the lossless ZOR point [26],[30].

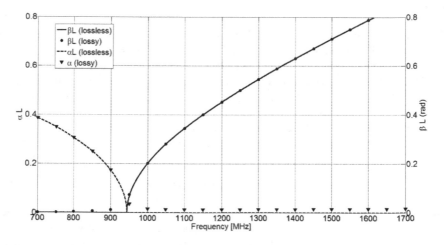

Figure 8. Dispersion diagram of lossless and lossy interdigital capacitor loaded CPW

5.2. Zeroth order resonator RFID antenna measurements

Again, the proposed ZOR RFID antenna with the capacitor loaded CPW is shown in Fig. 7. The antenna is composed of four series connected unit cells, where each unit cell has a layout similar to the image in Fig. 6. The proposed antenna has a 50 ohm CPW at one end and a high characteristic impedance short circuit line on the other end similar to [26] and [30]. The Higgs-2 by Alien Techonologies [29] RFID ASIC was used and attached at the port of the antenna (at the center). The Higgs-2 has an input impedance of Z_{in} = 13.73 + j142.8 Ω at 915 MHz. The antenna was designed on a Rogers TMM4 substrate with ε_r = 4.5, tan δ = 0.002 and a substrate thickness of H = 1.524 mm. The design parameters of the proposed ZOR RFID antennas are given in Table 1 and [30].

A wider central strip was used to obtain the required series capacitance as shown in Fig. 7 and the gap between the central conductor and reference conductors on either side was made as large as possible so that the parasitic shunt capacitance could be made as small as possible. This ensured a dominant series capacitance created by the interdigital capacitance and the input impedance of the passive UHF RFID ASIC connected to the antenna port. Furthermore, this will simplify the ABCD matrix representation of each unit cell.

The ZOR RFID antenna shown in Fig. 7 was simulated in Ansoft HFSS v.13. The simulated input resistance, reactance and reflection coefficient are shown in Fig. 9, Fig. 10 and Fig. 11, respectively. The fabricated prototoype ZOR RFID antenna is shown in Fig. 12 [30].

C	2.4 pF	w_3	0.66 mm
W	8.82 mm	S_1	12.17 mm
L	17.56 mm	S_2	0.35 mm
S	7.96 mm	L_1	5 mm
w_1	0.4 mm	l	16.2 mm
W_2	3 mm	g	0.36 mm

Table 1. Design parameters of proposed ZOR RFID antenna

Next, to measure the read range of the prototype tag, an Alien Technologies ALR-9900 RFID reader was used [29] (with maximum output power of 1W). It was connected to a circularly polarized antenna with a gain of 6dBi and the RFID Tag was placed in an anechoic chamber. A read range of 3.4 m was determined with the RFID reader; however the *max read range* was not determined because the overall dimensions of the anechoic chamber were too small. An alternate method has been provided in [30] and [31] to predict the maximum achievable read range based on system power levels and measurements. This method uses the Friis transmission equation and the fact that a certain minimum power is required to activate the tag. Using this information the output power of the RFID reader was reduced until the reader could no longer detected the tag at 3.4 m. The required attenuation was 7 dB. Then the following equations were used to predict the maximum read range:

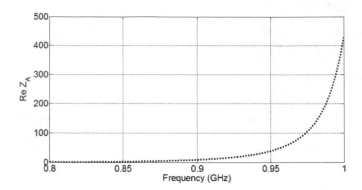

Figure 9. Proposed ZOR RFID antenna input resistance

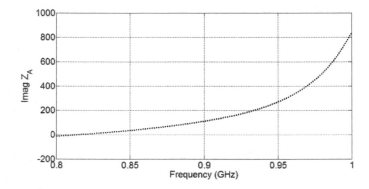

Figure 10. Proposed ZOR RFID antenna input reactance

$$P_{r_{min}} = \frac{P_{t_{max}} G_t G_r \lambda^2}{(4\pi R_{max})^2} \tag{36}$$

and

$$P_{r_{min}} = \frac{P_{t_{max}}}{\alpha} \frac{G_t G_r \lambda^2}{(4\pi R_{measured})^2} \tag{37}$$

Since (36) and (37) both use minimum received power, they can be equated to produce

$$R_{max} = 10^{\alpha_{dB}/20} R_{measured} \tag{38}$$

Putting $\alpha = 7$ dB and $R_{measured} = 3.4$ m in (38) gives a predicted max read range of 7.6 m which meets or exceeds the performance of similar and large passive UHF RFID tags available on the market today.

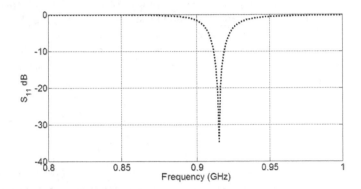

Figure 11. Input reflection coefficient of proposed ZOR RFID antenna

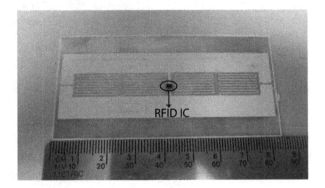

Figure 12. Fabricated ZOR RFID antenna [30]

6. Conclusion

The expanding use of passive UHF RFID systems has increased the performance demands on readers, tags, software and manufacturing costs. Because of these new constraints, the desire for more compact and better performing tags is beginning to grow. In this chapter, a summary of passive UHF RFID systems has been presented with several of the key antenna design requirements mentioned. Following this introduction, background on left-handed propagation, co-planar waveguides and interdigital capacitor loaded co-planar waveguides have been introduced and summarized. From these sections, the ZOR-RFID antenna for passive UHF RFID tags is presented. The operating principle behind the ZOR-RFID antenna is the use of interdigital capacitors along the length of the antenna to support wave propagation. Furthermore, the capacitive input impedance of the passive RFID ASIC attached to the port of the antenna supports propagation in a manner similar to the interdigital capacitors. This allows the ASIC to still harvest power and communicate while supporting wave propagation. Measurements show that a predicted 7.6 m read range is possible with this new antenna design. This read range is comparable to existing commercially available passive UHF RFID tags with similar overall sizes.

7. Future work

There are several different avenues of future work possible. The first topic of interest is to reduce the overall size of the ZOR RFID prototype antenna. This could be done by using resonator elements instead of the interdigital capacitors. Further development on printing the ZOR-RFID antenna on flexible substrates would be of great interest. Maybe the investigation of paper, LCP and Kapton substrates could be performed. Extending this work to develop a multi-band antenna would also be possible. This would allow this antenna design to be used in multiple countries.

Author details

Muhammad Mubeen Masud and Benjamin D. Braaten

North Dakota State University, Fargo, U.S.A.

References

[1] Finkenzeller, K. RFID Handbook: Fundamentals and Applications in Contactless Smart Cards and Identification. John Wiley and Sons, West Sussex, England; 2003.

[2] Rao, K.V.S., Nikitin P.V., Lam S.F. Antenna Design for UHF RFID Tags: A Review and a Practical Application. IEEE Transactions on Antennas and Propagation December 2005; 53(12) 3870-3876.

[3] Calabrese C., Marrocco G. Meander-slot antennas for sensor-RFID tags. IEEE Antennas and Wireless Propagation Letters 2008; 7 5-8.

[4] Amin Y., Batao S., Hallstedt J., Prokkola S., Tenhunen H., Zhen L.-R. Design and characterization of efficient flexible UHF RFID tag antennas: proceedings of 3rd European Conference on Antennas and Propagation, March 2009, Berlin, Germany.

[5] Stupf M., Mittra R., Yeo J., Mosig J. R. Some novel design for RFID antenna and their performance enhancement with metamaterials. Microwave and Optical Technology Letters February 2007; 49(4) 858-867.

[6] Braaten B.D., Scheeler R. P., Reich R. M., Nelson R. M., Bauer-Reich C., Glower J., Owen G. J. Compact Metamaterial Based UHF RFID Antennas: Deformed Omega and Split-Ring Resonator Structures. The Applied Computational Electromagnetics Society Journal 2010; 25(6) 530-542.

[7] Braaten B. D., Reich M., Glower J. A Compact Meander-Line UHF RFID Tag Antenna Loaded with Elements Found in Right/Left-Handed Coplanar Waveguide Structures. IEEE Antennas and Wireless Propagation Letters 2009; 8 1158-1161.

[8] Dacuna J., Pous R. Miniaturized UHF tags based on metamaterials geometries Building Radio Frequency Identification for the Global Environment, July 2007, http://www.bridge-project.eu.

[9] Sanada A., Caloz C., Itoh T. Characteristics of the composite right/left-handed transmission line. IEEE Microwave and Wireless Components Letters February 2004; 14(2) 68–70.

[10] Caloz C., Itoh T. Electromagnetic Metamaterials. Piscataway-Hoboken, NJ: Wiley-IEEE Press, 2005.

[11] Sanada A., Kimura M., Awai I., Caloz C., Itoh T. A planar zeroth-order resonator antenna using a left-handed transmission line: proceedings of 34th European Microwave Conference, October 2004, Amsterdam, Netherlands.

[12] Rennings A., Liebig T., Abielmona S., Caloz C., Waldow P. Triband and dual-polarized antenna based on composite right/left-handed transmission line: proceedings of 37th European Microwave Conference, October 2007, Munich, Germany.

[13] Lee C.-J., Leong K. M. K. H, Itoh T. Compact dual-band antenna using an anisotropic metamaterials: proceedings of 36th European Microwave Conference, September 2006, Manchester, United Kingdom.

[14] Kanan R., Azizi, A. UHF RFID transponders antenna design for metal and wood surfaces: proceedings of IEEE International Conference on RFID, April 2009, Orlando, Florida.

[15] Teco: RFID Smart Shelf: http://www.teco.edu/research/projects/smartshelf/. (accessed 13 August 2012).

[16] Stutzman, WL., Thiele, GA. Antenna Theory and Design 2nd edition: John Wiley and Sons, Inc., New York; (1998)

[17] Braaten B.D., Owen G. J., Vaselaar D., Nelson R. M., Bauer-Reich C., Glower J., Mor-lock B, Reich M., Reinholz A. A printed Rampart-line antenna with a dielectric super-strate for UHF RFID applications: in proceedings of the IEEE International Conference on RFID, April, 2008, Las Vegas, NV.

[18] Rao K. V. S., Nikitin P.V., Lam S.F. Antenna Design for UHF RFID Tags: A Review and a Practical Application: IEEE Transactions on Antennas and Propagation December 2005, 53(12) 3870-3876.

[19] Wen C.P. Coplanar Waveguide: A Surface Strip Transmission Line Suitable for Non-Reciprocal Gyromagnetic Device Application. IEEE Transaction on Microwave Theo-ry and Techniques December 1969; 17(12) 1087-1090.

[20] Pozar D. M. Microwave Engineering 3rd edition, John Wiley and Sons, Inc., Hobo-ken, New Jersey 2005.

[21] Collin R. E. Foundations for Microwave Engineering 2nd edition, John Wiley and Sons, Inc. Hoboken, New Jersey 2001.

[22] Velez A., Aznar F., Bonache J., Valazquez-Ahumada M. C., Martel J., Martin F. Open complementary split ring resonators (OCSRRs) and their application to wideband CPW band pass filter: IEEE Microwave and Wireless Components Letters April 2009; 19(4) 197-199.

[23] Eleftheriates G. V., Balmain K. G. Negative-Refraction Metamaterials: Fundamentals Principles and Applications, John Wiley and Sons, Hoboken, New Jersey 2005.

[24] Gupta KC., Garg R., Bahl I.J. Microstrip Lines and Slotlines, Artech House, Massa-chusetts 1979.

[25] Hilberg W. From Approximations to Exact Relations for Characteristic Impedances. IEEE Transaction on Microwave Theory and Techniques May 1969; 17(5) 259-265.

[26] Lai CP., Chiu SC., Li HJ., Chen SY. Zeroth Order Resonator Antennas Using Induc-tor-Loaded and Capacitor-Loaded CPWs. IEEE Transaction on Antennas and Propa-gations September 2011; 59(9) 3448-3453.

[27] Lai A., Itoh T., Caloz C. Composite right/left-handed transmission line metamateri-als. IEEE Microwave Magazine September 2004, 5(3) 34–50.

[28] Caloz C., Itoh T. Metamaterials for high-frequency electronics. Proceedings of the IEEE October 2005, 93(10) 1744–1752.

[29] Alien Technologies: http://www.alientechnologies.com. (accessed 25 August 2012).

[30] Masud M.M., Ijaz B. and Braaten B.D., "A Coplanar Capacitively Loaded Zeroth Or-der Resonator Antenna for Passive UHF RFID Tags," Submitted for review in the IEEE Transactions on Antennas and Propagation.

[31] Vaselaar D. Passive UHF RFID design and utilization in the livestock industry. Master's thesis. North Dakota State University, Fargo; 2008.

Chapter 7

Integrating RFID with IP Host Identities

Steffen Elmstrøm Holst Jensen and
Rune Hylsberg Jacobsen

Additional information is available at the end of the chapter

1. Introduction

The "Internet of Things" semantically means "a world-wide network of interconnected objects uniquely addressable, based on standard communication protocols" [1]. The vision describes a world that enables physical objects to act as nodes in a networked physical world [2]. The terms "Internet of Things" can be attributed to The Auto-ID Labs, a world-wide network of academic research laboratories in the field of networked RFID and emerging sensing technologies [2]. Together with EPCglobal®, these institutions have been architecting the Internet of Thing since their establishment. Their focus has primarily been on the development of the Electronic Product Code™ (EPC) to support the wide-spread use of RFID in modern, global trading networks, and to create an industry-driven set of global standards for the EPCglobal Network.

EPCglobal Network was created for "traditional" low-cost tags [3]. The main functionality of the EPCglobal Network is to provide data assigned to a specific tag, so that each RFID read event can be stored in a database and applications can be built on this data. Since tags were not originally considered to carry or compute additional data, the EPCglobal Network does not traditionally provide a mechanism to address remote tags from networked applications.

The data flow in these networks works from tags via readers to a couple of networked servers. Passive, low-cost RFID tags are widely available and the EPCglobal Network was defined to support open-loop supply chain applications. Basically, this is accomplished by allowing servers to communicate over the Internet. Although RFID technology is quite accepted in closed-loop applications, the evolution towards open-loop systems using the EPCglobal Network with distributed databases did not take place as predicted due to problems in the access control layer of such systems.

RFID-sensor networks are an emerging part of the Internet of Things [5]. These devices combine sensing capabilities with an RFID interface that allow the retrieval of sensed data. In fact, they can cooperate with RFID systems to better track the status of things e.g., their location, temperature, movements, etc. A sensor-enabled RFID tag (also known as sensor-tags) is an RFID tag which contains one or more sensors to monitor some physical parameter (e.g., temperature) but also contains the same identification function as a "normal" RFID tag does. This kind of sensor tag may fall into class 2, class 3 or class 4 in EPCglobal's tag classification [3]. A fully passive, class 2 sensor-tag can measure physical parameters, i.e., use sensors, only when powered by a reader. In contrast, class 3 tags are battery assisted. They can work independently of the reader and can be suitable for RFID-sensor networks.

In this chapter, we will discuss different ways to achieve the Internet of Things vision by internetworking passive RFID tags over IPv6. The chapter is organized as follows: Section 2 presents related works and discusses the novelty of the work presented here. Section 3 introduces the key technologies for the convergence of RFID and Internet namespaces and to provide an address mapping needed to internetwork passive RFID tags. In Section 4, some common examples of RFID usage are given and discussed in the context of globally networked tags. Subsequently, Section 5 introduces a testbed built to study the interconnection of passive RFID tags over IPv6. The different strategies that can be used for integrating RFID with IPv6 are discussed in Section 6 and this discussion is followed by mobility considerations in Section 7. Finally, Section 8 concludes the discussion and outlines anticipated future work in this area.

2. Related works

Most objects in our surrounding are not equipped with microprocessors and hence cannot attach to a computer network. However, these objects can be equipped with passive, low-cost RFID tags either as tags integrated or adhesively stuck to the object and hereby provide a mean of communications. Dominikus et al. [14] has suggested a way to integrate passive RFID systems into the Internet of Things, by using readers that function as IPv6 routers. In their work, an IPv6 addressing scheme that map tag IDs to network addresses was defined. Furthermore, the mobility problem, which arises when tags physically moves around, was investigated and the use of Mobile IPv6 (MIPv6) to cope with tag mobility was suggested. In contrast to the work presented by Dominikus et al. [14], this chapter opens the discussion on the proper formatting of the IPv6 addressing by introducing cryptographic hashing techniques as well as the possibility of separating identity and location information when forming an IPv6 address. The use of hashing techniques to construct an IPv6 address from an EPC, as opposed by using a compressed EPC format [14], eases practical implementations and allows the use of the same mapping scheme for all EPC types.

An alternative approach is to provide the tags themselves with the IPv6 protocol stack, making them able to use IPv6 communication over the Internet whenever close to a reader. This requires several changes to the design of existing tags. In this case, the tags do all the work themselves and need a separate power source. A solution where the tags are modified to hold the IPv6 stack on them is discussed by Rahman et al. [4]. The tags EPC, which is its identity, would then be made into a part of the tags IPv6 address due to the design of the tags proposed. This makes these tags too expensive for integration into the Internet of Things since the price of the tags could easily exceed the value of the "things" themselves.

Barish et al. [13], describes a somewhat similar setup than the one proposed here. In their approach, a global address manager is used to keep track of tags. The basic idea is that an application sends the EPC to a global server along with the IP address that the tag has been associated with. When a corresponding node wants to communicate with the tagged object, it contacts the last known address. If the tag is in the field of the reader the connection is established and communication can begin. If the tag is not present at the location the request is redirected to the global address server that returns the tag's present address or just redirects the request to the correct address. In contrast to the proposed solution by Barish et al. [13], the approach described here does not include extra nodes in the network to construct network addresses but adds functionality to the RFID readers residing at the network edge.

Xu et al. [25] proposed a general address mapping scheme based on a proprietary protocol named General Identity Protocol (GIP). The scheme takes all existing RFID systems into account, and allows heterogeneous RFID systems to interwork over the Internet. This is accomplished by mapping RFID tag identifiers to IPv6 addresses, constructing a GIP message with details of the RFID systems in use, and finally encapsulating the message in IPv6 and routing the packet over the Internet. This chapter describes a solution that minimizes the need for control protocols.

3. Enabling technologies

There are a couple of ways to interconnect objects by using RFID with IPv6 [6]. One solution would be to give the tags the ability to communicate via the Internet. The communication can be both reader-initiated and tag-initiated. The latter requires specific tags that require electrical and processing power to be available in the tag such as e.g., EPC class 3 tags. Most of the computational work takes place in the tags, i.e., the tag is reachable and visible as an IPv6 connected host as long as it is within the electric field of a reader.

Passive RFID tags, such as EPC class 2 tags, do not have the possibility to power a network protocol stack and therefore a network address cannot be directly assigned to the tag's microchip. However, the passive tag can be represented by virtual interfaces residing in the reader interrogating the tag.

3.1. Radio Frequency Identification (RFID)

RFID systems are composed of one or more readers and several electronic tags. Tags are characterized by a unique identifier that takes the form as a binary number. They are applied to objects and even persons or animals as implants. From a physical point of view, an

RFID tag is a small microchip attached to an antenna that is used for both receiving the reader signal and transmitting the tag ID. The dimensions of each tag can be very small with tag dimensions down to 0.05 mm x 0.05 mm with a thickness of 0.005 mm [7]. There are more than 60 tag manufacturers world-wide [8].

RFID tags will act as electronic identification for physical objects to which they are linked. In the Internet of Things, all objects, virtual as well as physical, are interconnected and reachable via for example IPv6 in combination with RFID technology [6]. Essentially, the tag connects to physical objects that we want to authenticate and track when they come in contact with readers. A reader can read or modify tag's information. The back-end database keeps information related to different tags/readers.

For reader-initiated communication the reader triggers the tag's transmission by generating an appropriate signal, which represents a query for the possible presence of tags in the surrounding area and for the reception of their identification codes (IDs).

Active tags come with a power source that can drive a microprocessor (or microcontroller). Furthermore, it allows a stronger electromagnetic field to be generated in response to an incoming RFID air protocol message and larger read distances can be achieved. More advanced active tags or sensor-tags may run additional software and can be equipped with communication software such as an IP protocol stack [9].

In contrast, passive tags rely on the incoming electromagnetic field from the reader to power the circuit and to deliver power to drive the response to an interrogating request. These devices do not run communication software and cannot be actively involved in a protocol message exchange. To communicate with these devices there is a need for software agents to act on their behalf.

RFID tags can only be "online" when they are in the electric field of a reader field. For high velocity applications, where tags only remain certain seconds in a reader field, the proposed approach of networking these tags is not applicable.

3.2. RFID namespaces

Essentially, RFID comes with two namespace to be used with RFID applications: the EPC addresses and the Object Name Service (ONS). A namespace can represent objects as well as concepts and may be generalized as a container for a set of identifiers (names). The EPC is an identifier based on the standards established by EPCglobal® [10]. It is designed to allow the automatic identification of objects anywhere. EPC defines three layers of identity: the *pure* identity, the encoding layer identity and the physical realization of an encoding. The EPC tag data standard [10] identify how existing coding systems such as the GS1 family codes for serialized human readable representations e.g. GTIN, GCN, SSCC, GRAI, GIAI, GSRM, GDTI and a small number of other identities should be embedded within the EPC.

A canonical representation of an EPC is the *pure identity* Uniform Resource Indicator (URI) representation, which is intended for communicating and storing EPC in information systems, databases and applications. The purpose is to insulate EPCs from knowledge about the physical nature of the tag, so that although 64-bit tags may differ from 96-bit tags in the choice of binary header values and in the number of bits allocated to each element or field within the EPC, the pure identity URI format does not require the information system to know about these details. Hence, the pure identity URI can be regarded as a pure identifier [10]. Tags are identified by URIs such as e.g., urn:epc:id:sgtin:0523141.000024.120 that comprise both tag number and associated coding scheme.

Encoding is the process of translating the pure identity EPC into a specific instantiation incorporated into tags for a specific purpose. During the encoding process the URI information is translated into a binary encoding that is stored in the tag. Subsequently, translating between the different levels of representation can be accomplished in a consistent way.

Figure 1 shows the structure of the data layout of an EPC code for the Global Trade Item Number (GTIN) and Serialized Global Trade Item Number 96-bit (SGTIN-96) tag.

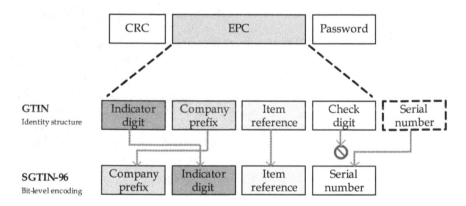

Figure 1. Tag data layout example.

EPC generation 1 standards, i.e., class 0 and class 1 tags, use a Cyclic Redundancy Check (CRC) to verify data integrity and a password as a "kill code" to disable the tags. The password must never be transmitted under any circumstances [3]. The *Item reference* identifies a class of objects to be tagged and it allows the grouping of items. The *Serial number* identifies an instance of a particular item. Company prefixes (also known as General Manager Numbers) point to the organization responsible for the subsequent partition. Finally, *Indicator digits* are used to specify length, type, structure, version, and generation of the EPC. This latter part is further used to guarantee uniqueness in the EPC namespace. For the GTIN encoding a *Check digit* is used.

Since the EPC is the only required data stored on a tag, it must be used as a "pointer" to find additional data about an object to which it attaches. This additional data should be stored on a server connected to the enterprise network or to the Internet. The server is identified via a look-up system which is called ONS.

ONS acts as a directory service for organizations wishing to look up product numbers (also known as EPC numbers) on the Internet. The ONS is operated as part of the EPCglobal Network. It is based on the well-known DNS service and it realizes the link between EPC numbers and EPC Information Services (EPCIS) as illustrated in Figure 2. When an RFID reader reads a tag, the EPC is passed to a middleware which then looks the EPC up either on the local machine, or enquires ONS through the Internet.

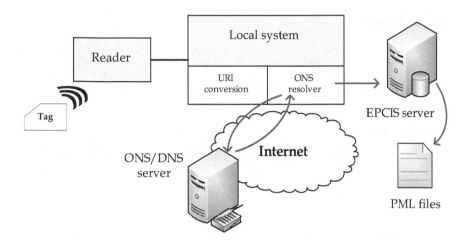

Figure 2. The Object Name System (ONS). Adapted from [11].

The ONS resolution process takes the EPC code and returns network location(s) where information resides, i.e., the EPCIS server, which typically holds web pages with information about tags. The Physical Markup Language (PML), based on XML technology, is intended to be the standard in which information about tags should be written.

In contrast, the DNS of the Internet will handle many more requests in the future. Therefore, enterprises will likely maintain ONS servers locally, which will store information for quick retrieval. Hence, a manufacturer may store ONS data from its current suppliers on its own network, rather than pulling the information off a Web site every time a shipment arrives at the assembly plant.

3.3. Internet namespaces

There are two principal namespaces in use in the Internet: IP addresses and domain names. Domain names provide hierarchically assigned names for some computing platforms and some services. Each level in the hierarchy is delegated from the level above. Email, Hypertext Transfer Protocol (HTTP), and Session Initiation Protocol (SIP) addresses all reference domain names to mention its most wide-spread use.

On the network layer, IP addresses are used. IPv6 was introduced in the 90'ies due to the foreseen lack of globally unique IPv4 addresses, resulting in a protocol specification released in 1998 [17]. The IPv6 address is a 128-bit address that takes the form of a 64-bit network prefix appended by a 64-bit host suffix/interface identifier. The network prefix is used for routing purpose and determines the location of the host in the Internet. The host itself is identified by an interface ID. Figure 3 shows the IPv6 address format and gives an example on how a 96-bit EPC can be mapped to a network address.

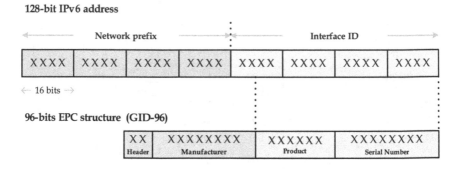

Figure 3. IPv6 address format compared to a 96-bit General Identifier (GID-96) EPC format. An 'X' indicates a grouping of 4 bits.

It can be observed form the figure that not all 96 bits of the EPC can be fitted within the host suffix/interface identifier of an IPv6 address. Because of this deficiency, the implementers of an RFID-to-IPv6 mapping scheme is faced with a number of design options. These options basically govern the strategy for the mapping and are the subject of our discussion in Section 6.

3.4. Cryptographically generated addresses

Cryptographically Generated Addresses (CGAs) are IPv6 addresses for which the host suffix/interface identifier is generated by computing a cryptographic one-way hash function from a binary input such as e.g., a public key [27]. CGAs are intended to be globally unique in a statistical sense but these may not necessarily be routable addresses at the IP layer [12].

The Overlay Routable Cryptographic Hash Identifiers (ORCHID) is a new, experimental class of identifiers based on CGAs. ORCHIDs have an IPv6-like address format and can be used with existing applications built on IPv6 [12]. These identifiers are intended to be used as pure endpoint identifiers for applications and Application Programming Interfaces (APIs) and not as identifiers for network location. This is in contrast to the IPv6 address that uses the 64-bit network prefix as locator [17].

While ORCHIDs use public cryptographic keys as input bit strings, it is possible to use the binary EPC encoding instead. The algorithm to generate an ORCHID in an RFID context is outlined below [12]. The algorithm takes a bitstring and some *context identifier* as inputs and produces an ORCHID output that is formatted as an IPv6 address.

$$Input := anybitstring \tag{1}$$

$$HashInput := ContextID \mid Input \tag{2}$$

$$Hash := Hash\_function(HashInput) \tag{3}$$

$$ORCHID := Prefix \mid Encode_n(Hash) \tag{4}$$

Concatenation of bitstrings is denoted '|'. The *Input* is a bitstring that is unique within a given context. The *Context ID* is a randomly generated value defining the expected usage context for the particular ORCHID and the hash function to be used for generation of ORCHID in this context. The purpose of a context ID is to be able to differentiate between various experiments that share the ORCHID namespace. The *Hash_function* is a one-way hash function to be used to generate ORCHIDs such as SHA1 [23] or MD5 [24]. SHA1 and MD5 produce a 160-bit and a 128-bit output, respectively. *Encode_n* is a function to extract an n-bit-long bitstring from its argument. Finally, *Prefix* is an IPv6 network prefix.

To construct a CGA an input bitstring and context identifier are concatenated to form an input datum, which is then fed to the cryptographic hash function. The result of the hash function is processed by an encoding function, resulting in an n-bit-long output. This value is prepended with the network prefix resulting in a 128-bit-long bitstring identifier that can be used for programming with the IPv6 API.

To create a CGA namespace for RFID tags the EPC of a tag and the network prefix assigned to the reader that interrogates the tag are used as input. Furthermore, an $Encode_{64}$ function is used to extract 64 bits from the hash. A key advantage of using hash values over the actual raw host identity resulting from the EPC is its fixed length. This makes protocol implementations easier and it alleviates the management of packet sizes. However, a claimed drawback is that CGAs work one-way, meaning that it is not possible directly to create the original identity from the hash.

A CGA can be globally unique or globally unique in a statistical sense. That is, the probability of the same CGA being used to refer to different entities in the Internet must be sufficiently low so that it can be ignored for all practical purposes. Even though CGA collisions are expected to be extremely rare, collisions may still happen since it is possible that two different input bitstrings within the same context may map to the same CGA. A second type of collision can happen if two input bitstrings, used in different contexts, map to the same CGA. In this case, the main confusion is about which context to use. In order to preserve a low enough probability of collisions, it is required that applications ensure that distinct input bitstrings are either unique or statistically unique within a given context. By adhering to the EPCglobal standards, this requirement is fulfilled.

3.5. Host identities and host identity protocol

A host identity is an abstract concept assigned to a computing identity platform. In this section, we will generalize this concept to cover thin compact platforms that can be equipped with RFID tags. The discussion starts by introducing the host identities and the host identity protocol [15][16].

The Host Identity Protocol (HIP) supports an architecture that decouples the transport layer (TCP, UDP, etc.) from the internetworking layer (IPv4 and IPv6) by using public/private key pairs, instead of IP addresses, as host identities [15][16]. The public keys are typically, but not necessarily, self-generated. HIP introduces a new Host Identity (HI) namespace, based on these public keys, from which end-point identifiers are taken. Host identifiers are used to bind to higher layer protocols instead of binding to IP addresses. A key benefit of this approach is that it is compatible with existing APIs such as the socket API. HIP uses existing IP addressing and forwarding for locators and packet delivery, respectively.

Figure 4 illustrates the difference between binding of the logical entities service and endpoints to an IP address (left side of figure). The service typically binds to the IP stack via the socket API. By using the host identity abstraction of the HIP architecture, the service and the end-point bind to the host identity whereas the location is still anchored with the IP address.

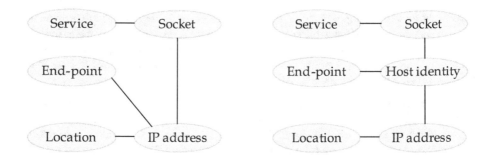

Figure 4. Illustration of the difference between the bindings of the logical entities. Adapted from [15].

There are two main representations of the host identity, the full Host Identifier (HI) and the Host Identity Tag (HIT). The HI is a public key and directly represents the identity. The HIT is the operational representation of a host. It has a 128-bit long representation and is used in the HIP payloads to index the corresponding state of the end hosts. By introducing an identity concept at the network layer, where every host is represented by an asymmetric key pair consisting of a public and private key, it turns IP addresses into pure locators.

The proposed HI namespace fills an important gap between the IP and DNS namespaces. A public key is used as the HIP Host Identity (HI), while the private key serves as proof of ownership of the public key. To seamlessly integrate HIP with protocols above the network layer, a 128-bit cryptographic hash of the HI, i.e., the HIT, was introduced to fit the IPv6 address space. The HIT is a statistically unique flat identifier. When HIP is used, the transport layer binds to HITs. In this process it becomes unaware of the IP addresses that are used for routing.

To be able to setup communication between peers that use HI, a light-weighted protocol exchange called the HIP Base Exchange has been specified. In Figure 5, the HIP Base Exchange is adapted to an RFID setup. The setup is somewhat similar to the one presented by Urien et al. [11].

Since the deployed tags are passive, there is a need for a proxy to act on behalf of the tags in the protocol exchange. The role of this proxy will be explained further in Section 5.

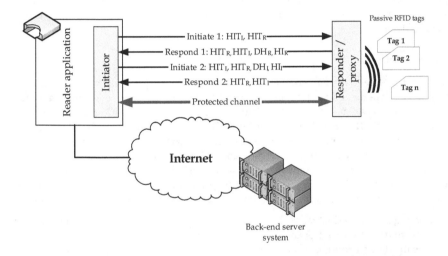

Figure 5. HIP Base Exchange adapted to a RFID communication scenario. Adapted from [15].

The HIP Base Exchange is a four-way handshake between two hosts wanting to initiate communication (see Figure 4). The *Initiate 1* packet is the first packet sent in the handshake. It is an unencrypted and unsigned packet, meaning that the Initiator would like to talk HIP with the Responder. The HIP packet contains the HIT of the Initiator (HIT_I) and the Responder (HIT_R). The responder's IP address can be derived from the DNS. *Respond 1* is sent as a reply to the *Initiate 1* packet. Besides the HIT_I-HIT_R identity pair, it contains a cryptographic puzzle challenge, and Diffie-Hellman parameters (DH_R) for the Diffie-Hellman key agreement. The Diffie–Hellman key exchange method allows two parties that have no prior knowledge of each other to jointly establish a shared secret key over an insecure communication channel. Subsequently, the secret key can be used for encryption and integrity protection of the communication channel. The purpose of the HIP puzzle mechanism is to protect the Responder from denial-of-service attacks. The *Initiate 2* packet returns the corresponding Diffie-Hellman parameter (DH_I) to the Responder. It carries an encoded solution to the puzzle.

Upon reception of the *Initiate 2* packet, the responder can now generate the keying material, and it is capable of using it in encryption and integrity protection algorithms. The Response 2 packet completes the HIP Base Exchange. After the Base Exchange, there is no longer difference between the Initiator and Responder and data can securely be exchanged between the communicating peers.

4. Use cases and application examples

RFID applications are numerous and far reaching [8]. The most interesting and widely used applications include those for security and access control, supply chain management, and the tracking of important objects and personnel. This section outlines a number of commonly encountered use cases for RFID technology, and discusses these in the context of networked RFID tags.

4.1. Access control

Access control systems are an important part of the security of government buildings, companies, schools, residences and private areas and RFID technology has been widely adopted in access control systems. These systems often use RFID identification cards based on the IEC/ISO 14443 [18], IEC/ISO 15693 [19], or IEC/ISO 18000 standards [20]. The identification cards work much like a traditional key for unlocking doors or otherwise granting access. However, RFID technology does not provide authentication to the holder of the RFID card (or tag). Any unauthorized people holding an authorized RFID card could get access to secured area. Therefore, RFID technology should be combined with other means of identification such as e.g., face recognition to strengthen the security of the access control system.

By associating a passive RFID tag such as a key card with a globally unique IPv6 address we will be able to use access control and security policy mechanisms with Internet technologies to provide the desired access control applications. In this scenario a door locking mechanism would be connected over the Internet resulting in a more open system architecture.

4.2. Supply chain management

Most supply chain applications involve the concept of inventory tracking. An example of a proposed use of RFID is to ensure safety in the supply chain [21].

To illustrate the potential of using network RFID tags with supply chain applications an example taken from the Tag Data Standard v1.6 issue 2 [10]. The example text is quoted below:

"... a shipment arriving on a pallet may consist of a number of cases tagged with SGTIN identifiers and a returnable pallet identified by a GRAI identifier but also carrying an SSCC identifier to identify the shipment as a whole. If a portal reader at a dock door simply returns a number of binary EPCs, it is helpful to have translation software which can automatically detect which binary values correspond to which coding scheme, rather than requiring that the coding scheme and inbound representation are specified in addition to the input value."

Each of the cases tagged will be given a unique IPv6 address when they enter the electric field of a reader. This process involves the extracting of the essential bitstring of the SGTIN identifier for each case. Likewise, the returnable pallet and the shipment as a whole will be given IPv6 addresses that can be built based on the GRAI and the SSCC, respectively. By using the assigned IPv6 unicast addresses it is possible to establish communication to individual cases as well as the pallet. However, it may be of less interest to address individual cases at this point in the supply chain but rather to address the ensemble of cases. By introducing multicasting at the network layer it can be possible to communicate with groups of cases on the pallet.

4.3. Object/asset tracking

Because moving objects can easily carry RFID tags, a common use is to track the movement of people and the information associated with them. By associating a particular tag's EPC with a global network address the task of tracking the object/asset become equivalent to locating a mobile host in the network. In general, this is a key challenge in mobility research and several solutions have been proposed [22][26], and this will be the subject of our discussion in Section 7. Another interesting use case can be applied to sensor-tags. When these sensor-tags connect to a network sensor data can be retrieved from the tag.

5. Networked RFID testbed

To study the internetworking of objects with passive RFID tags, a simple testbed has been built. The approach makes use of an RFID reader and an application that works as a proxy for the tags we wish to communicate with. The proxy is capable of making a virtual representation of the passive RFID tag on the Internet by creating a Virtual Network Interface (VNI) with an IPv6 address that can be attributed to each tag that comes within the electric field of a reader. Hence, the tags do not terminate IPv6 traffic directly but merely communicate with an entity which represents the tag (physical object) that we wish to communicate with.

The approach taken is software-oriented. The application runs as standalone but it can be embedded on the reader or it can be run on a computer local to the reader. The application receives the EPC of a tag attached to a physical object via the reader. The application then creates an IPv6 address from one of the mentioned methods. Hereafter, it is possible to route IP traffic to the particular Internet end-point. This will in effect make the application act as a proxy that for example can keep the most recently read tags "online". The solution gives a one-to-one mapping of physical objects to the virtual representations that are needed to communicate over the Internet.

Figure 6 illustrates the system implemented. In practice, the application host has a predetermined number of Virtual Network Interfaces (VNIs) installed. These interfaces work as the online virtual representation of the tag swiped at the reader. In other words, this is the interface the outside world can contact. In the testbed, the network interfaces are virtualized in a way similar to a loopback interface [17]. As the system works as a testbed the database is merely there as a logging service. In the future, it is planned to use the database as foundation for a local ONS. The corresponding node is there to illustrate possible communication over the Internet.

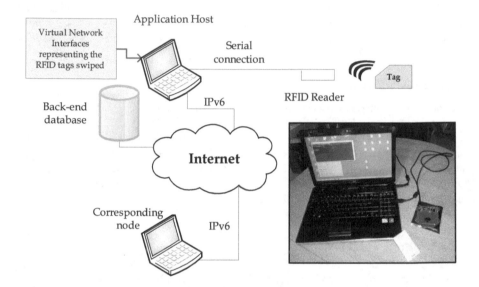

Figure 6. Simple setup to give RFID tags virtual identification on the Internet. A RedBee RFID Reader v1.1 is used. The application is built on the Microsoft®.NET connection software.

Figure 7 shows a state machine diagram for a single VNI resulting from a tag swiped in an access control application.

When the tag is swiped at the reader, the application host creates an IPv6 address by combining the network prefix configured at the reader with the tags identity as illustrated by the Example in Table 1. In the initial state, the software is waiting for a TagSwipe event to occur. Subsequently, the interface is put online with the address constructed, and it is kept alive as long the *expiration time* is greater than 0 (zero) seconds.

Tags are only reachable while they are within reader range. This makes it hard to communicate with the real tag, simply because it is only reachable for a short duration of time. When the tag's attachment to the network is virtualized it is possible to set up an expiration value. This value effectively serves as the time the tags virtual representation on the network can be reached.

The tag identity together with the constructed IPv6 address and a timestamp is stored on the database. Table 1 shows an example of the steps taken to construct an IPv6 address from an EM4100 tag ID.

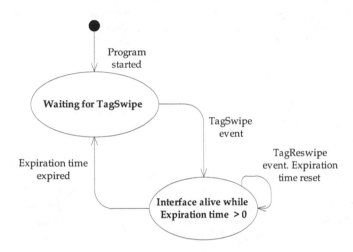

Figure 7. State machine for the virtual interface resulting from a TagSwipe event at the reader.

Tag identity (Example with EM4100 tag)	5 decimal numbers (40 bits)	127 0 58 207 19
Binary ID representation with left-zero-padding	64 bits	0000 0000 0000 0000 0000 0000 0111 1111 0000 0000 0011 1010 1100 1111 0001 0011
Converted to hexadecimal	4 groups of 4 hex. digits	0000 007f 003a cf13
Network prefix of RFID reader (example)	4 groups of 4 hex. digits	2001:16d8:dd92:aaaa::/64
Unicast IPv6 address associated with the tag	8 groups of 4 hex. digits	2001:16d8:dd92:aaaa::007f:003a:cf13/128

Table 1. Example of IPv6 network address construction from on EM4100 tag ID.

The application has no visual interface and all configurations must be done in software. For example, it is possible to use more than one virtual interface to represent the tags online. These interfaces need to be preinstalled, as already mentioned, and some parameters in the application need to be configured. Hereafter, it is possible to make use of at least 5 virtual interfaces.

Although focus is on the assignment of IPv6 unicast addresses, tags can also be assigned to become member of multicast groups thereby facilitating one-to-many communication. As an example an application may want to address all tags at a particular reader. Likewise, readers can become members of multicast groups hereby enabling communication to all readers in the multicast group e.g., a particular logical area. Details on how to operate an RFID in IPv6 multicast networks are beyond the scope of this chapter.

6. Strategies for interworking IPv6 with RFID tags

In this Section we will discuss different methods of mapping between an RFID namespace and an Internet namespace.

6.1. Address mappings

The most simple approach to find IPv6 addresses for tags is the mapping of the tag ID to an IPv6 address, i.e., the bits from the ID are used to form the IP address [14]. As different passive RFID standards exist there is no common ID structure for tags. Even within standards, there are different types of IDs with different structures. This means, that a general concept to map tag IDs to IPv6 addresses will not work.

Table 2 shows a list of some commonly encountered passive tags and their ID formats.

IC type	Frequency band	Memory	Standards compliance
EM4100 series	LF (125 KHz)	64 bits	EM4100. Proprietary standard issued by EM Microelectronics [28].
EM4450/4550	LF (125 KHz)	1024 bits	EM4450/4550. Proprietary standard issued by EM Microelectronics [29].
NXP Hitag family (Hitag 1, Hitag 2, Hitag S, Hitag μ)	LF (125 KHz)	256 bit to 2048 bit	IEC/ISO 18000-2 (Hitag μ) [20].
NXP Mifare family (Ultralight/ MF1S20/MF1S50/MF1S70/ DESFire EV1)	HF (13,56 MHz)	64 bytes to 4096 bytes 2K/4K/8K	ISO 14443A [18].
LEGIC Advant family (ATC128, ATC256, ATC1024, ATC2048, ATC4096)	HF (13,56 MHz)	128 bytes to 4096 bytes	IEC/ISO 15693 [19] (ATC128, ATC256, ATC1024), IEC/ISO 14443 A [18] (ATC 2048, ATC 4096).
NXP UCode HSL	UHF (868 MHz or 915 MHz)	2048 bit	ISO18000-4 and 18000-6B [20].
NXP UCode EPC Gen2	UHF (868 MHz or 915 MHz)	512 bit	EPCglobal class 1 gen2 and ISO 18000-6C [20].

Table 2. Common passive RFID tag and their characteristics.

An EPC with the length of 64 bits maps well in the IPv6 address format and can result in globally unique addresses. With longer EPCs it is impossible to map the EPC directly into the IPv6 address space and here specialized functions are needed. One solution would be to simply hash the longer EPC's into a length of 64 bits and then use the direct mapping method again. The hashing technique used to derive identifiers was described in Section 3.4, when the CGA namespace was introduced. Another method would be to identify if there are some bits in the longer EPC's that can be removed without affecting the uniqueness property of the tags.

A key benefit of the proposed solution is that there is no need to change the design of existing RFID technology with its EPC namespace conventions. The application can be installed on a computer connected to the reader, and then all objects with RFID tags that pass this reader will put the objects online and thereby giving them the ability to communicate over the Internet as long as the tag is within range of a reader.

	Strategy/method	Comments
Tag with ID of 64 bits or less	1. Use zero padding left-to-right to create a 64-bit input datum from the tag ID. Map this input to the host suffix/interface ID of the IPv6 address.	The tag ID can be read directly from the IPv6 address as the last 64 bit, i.e., the tag identification works two-ways. There is a risk of address collision with other networked systems that use IPv6 in the same logical subnet e.g., hosts using autoconfiguration. Uniqueness of the RFID naming space in use is conserved.
	2. Use CGA namespace adapted to RFID as described in Section 3.4.	Statistical uniqueness is achieved. The tag ID is hashed and cannot be directly read out of the IPv6 address. The reverse process of finding the tag ID from the IPv6 address requires a separate system to perform the mapping since the hashing works only one-way. Extra computational power is required because the method is based on cryptography.
Tag with ID of more than 64 bits	3. Use the tag ID as input bitstring to create a HIT. The ID should be treated as a public key in accordance with the HIP specification (see Section 3.5). End result is a 128-bit address compliant with the IPv6 addressing format.	Non-compliance with IPv6 address format. Addresses are non-routable because of the separation between locator and the end-point identifier. The solution requires changes in the reader's TCP/IP protocol stack. The solution will function with IPv6 API deployed in applications on the Internet.

Table 3. Overview of strategies for mapping Tag ID codes to IPv6 network addresses.

Table 3 outlines the different strategies for mapping of tag IDs to IPv6 addresses. Essentially, these divide into methods that work with tags of 64-bit identification or less and tags that use more that 64-bit for identification.

7. Mobility considerations

One of the largest challenges for a dynamic, networked system lies within the mobility support of the network. In the case described here, we consider a system of fixed readers that are connected in a common network infrastructure. Mobility arises when tags are moved between readers. Readers will be wired or wireless and they will have different communication ranges according to their MAC technology. Moreover, they will forward the read tag IDs to the server through the common network infrastructure.

When a tag moves from one reader to another, the network prefix will change but the host suffix/interface ID will still match the tag's EPC. The tag will in effect change its network address every time it passes a new reader. Hence, the challenge is to effectively keep track of tags when the address changes this rapidly.

There are basically two distinct ways to solve the mobility problem. One is a centralized approach, such as mobile IPv6 [30], where a central server, i.e., the home agent, is used to keep track of the mobile hosts that move around in the world. The mobile IPv6 architecture relies on the concept of a home agent and a care-of address. The method is based on some software on the network layer that can send messages to the home agent making sure that the home agent is holding an updated address list at all times. Initially, traffic destined to the mobile host is routed to the home network and subsequently tunneled to the foreign network that the host is visiting. Fortunately, IPv6 supports mechanisms to circumvent the triangular routing problem that arises in this setup [30].

Dominikus et al. [14], proposed to use mobile IPv6 to handle the mobility of IPv6-enabled tags. In their approach, the care-of address refers to the subnet of the RFID reader, where the tag is currently present. Whilst the care-of address is a globally unique address assigned to the host, i.e., the tag visiting a foreign network, the home agent address is specific to the enterprise using the issued tags.

Alternatively, mobility support can be obtained in a more distributed way by separating location and identity information. This can be achieved by using the HIP approach [22]. In this approach, there is a need to compute the routable IPv6 address from the given non-routable HIT the host has been given.

HIP allows consenting hosts to securely establish and maintain shared IP-layer state, allowing separation of the identifier and locator roles of IP addresses, thereby enabling continuity of communications across IP address changes. A consequence of such a decoupling is that new solutions to network-layer mobility and host multi-homing are possible [22].

8. Conclusions

Metcalfe's law states that the value of a telecommunications network is proportional to the square of the number of connected users of the system. When the law is applied to a network of objects on the scale predicted by the vision of the Internet of Things it is clear that a single, open architecture for networking physical objects is much more valuable than small scale and fragmented alternatives.

RFID plays an increasingly important in our daily life from management of goods, e-tickets, healthcare, transports, even the identity cards are embedded with RFID tags. In this chapter, we have sketched methods on how to use RFID technology to connect "things" over the Internet by using IPv6. This includes a discussion on the different strategies for mapping of tag IDs to globally unique IPv6 addresses.

For tags with large identification numbers (more than 64 bits) it is proposed to use cryptographic techniques to extract the 64 bits and use these to create a host suffix that is statistically unique.

A testbed used to experiment with the internetworking of low-cost, passive RFID tags to the Internet has been presented. Since these tags do not have electrical and processing power to run an IP protocol stack a virtual network interface (VNI) concept has been introduced. Proxies can be deployed on the edge of the Internet to act on behalf of these passive tags in a protocol message exchange.

To solve the mobility problem, two approaches have been discussed: one being the mobile IPv6 approach and the other being the HIP approach. Both have strengths and both have weaknesses. Mobile IPv6 will need some software to make the connection between the tags and the home agent. The HIP approach needs some computation to take place in order to be able to construct routable IPv6 addresses. Both approaches imply changes to be made to the Internet, as we know it today, before it is possible to effectively achieve the desired results.

Most RFID applications today include mobility as an essential part of their value creation. Therefore, future research in this area must focus on mobility aspects of the Internet of Things.

Author details

Steffen Elmstrøm Holst Jensen and Rune Hylsberg Jacobsen

Aarhus University School of Engineering, Denmark

References

[1] L. Atzori, A. Iera, and G. Morabito, "The Internet of Things: A survey," Computer Networks, vol. 54, no. 15, pp. 2787-2805, October 2010.

[2] S. Sarma, D.L. Brock, and K. Ashton, "The Networked Physical World, Proposals for Engineering the Next Generation of Computing, Commerce & Automatic-Identifica-

tion," MIT Auto-ID Center Massachusetts Institute of Technology, October 2000. Available from: *http://www.autoidlabs.org/single-view/dir/article/6/93/page.html*

[3]   B. Glover and H. Bhatt, "RFID Essentials," O'Reilly, 2006.

[4]   F. L. Rahman, M.B.I. Reaz, M.A.M. Ali, 2010, "Beyond the Wifi: introducing RFID system using IPv6," Proceedings of ITU-T Kaleidoscobe 2010, pp.1-4. Available from: *http://www.itu.int/pub/T-PROC-KALEI-2010*

[5]   L. Ho, M. Moh, Z. Walker, T. Hamada, and C.-F. Su, "A Prototype on RFID and Sensor Networks for Elder Healthcare: Progress Report," Proceedings of the 2005 ACM SIGCOMM Workshop on Experimental approaches to Wireless Network Design and Analysis (E-WIND'05), pp. 70-75, August 2005.

[6]   R.H. Jacobsen, Q. Zhang, and T.S. Toftegaard, "Internetworking Objects with RFID," in Deploying RFID - Challenges, Solutions, and Open Issues, Cristina Turcu (Ed.), ISBN: 978-953-307-380-4, InTech. Available from: *http://www.intechopen.com/books/deploying-rfid-challenges-solutions-and-open-issues/internetworking-objects-with-rfid*

[7]   T. Hornyak, "RFID powder", Scientific American, 298(2), 68-71, February 2008.

[8]   IDTechEx, "The RFID Knowledgebase." Available from: *http://www.idtechex.com/knowledgebase/en/nologon.asp*

[9]   J.J. Echevarria, J. Ruiz-de-Garibay, J. Legarda, M. Álvarez, A. Ayerbe, and J.I. Vazquez, "WebTag: Web Browsing into Sensor Tags over NFC," Sensors, vol. 12, pp. 8675-8690, 2012.

[10] GS1 EPC Tag Data Standard 1.6, September 2011. Available from: *http://www.gs1.org/gsmp/kc/epcglobal/tds*

[11] P. Urien, H. Chabanne, M. Bouet, D.O. De Cunha, V. Guyot, G. Pujolle, P. Paradinas, E. Gressier, and J.-F. Susini, "HIP-based RFID Networking Architecture," IFIP International Conference on Wireless and Optical Communications Networks (WOCN'07), pp. 1-5, July 2007.

[12] P. Nikander, J. Laganier, and F. Dupont, "An IPv6 Prefix for Overlay Routable Cryptographic Hash Identifiers (ORCHID)," Internet Society RFC 4843, April 2009. Available from: *http://tools.ietf.org/html/rfc4843*

[13] M. Barisch and A. Matos, "Integrating user identity management systems with the host identity protocol." 2009 IEEE Symposium on Computers and Communications, ISCC, pp. 830-836, 2009.

[14] S. Dominikus, M. Aigner, and S. Kraxberger, "Passive RFID Technology for the Internet of Things," 2010 International Conference for Internet Technology and Secured Transactions (ICITST), pp. 1-8, November 2010.

[15] R. Moskowitz and P. Nikander, "Host Identity Protocol (HIP) Architecture," Internet Society, RFC 4423. Available from: *http://datatracker.ietf.org/doc/rfc4423/*

[16] F. Al-Shraideh, "Host Identity Protocol," International Conference on Networking, International Conference on Systems and International Conference on Mobile Communications and Learning Technologies, ICN/ICONS/MCL 2006., pp. 203, 23-29 April 2006.

[17] S. Deering and R. Hinden, "Internet Protocol, Version 6 (IPv6) Specification," Internet Society, RFC 2460, December 1998. Available from: *http://tools.ietf.org/html/rfc2460*

[18] "ISO/IEC 14443 Identification cards - Contactless integrated circuit(s) cards - Proximity cards." Available from: *http://wg8.de/sd1.html#14443*

[19] "ISO/IEC 15693 Identification cards - Contactless integrated circuit(s) cards - Vicinity cards." Available from: *http://wg8.de/sd1.html#15693*

[20] "ISO/IEC 18000 Information technology - Radio frequency identification for item management." Available from: *http://www.iso.org/iso/home/store.htm*

[21] Bose and R. Pal, "Auto-ID: managing anything, anywhere, anytime in the supply chain," Communications of the ACM, vol. 48, no. 8, pp. 100-106, August 2005.

[22] P. Nikander, T. Henderson, C. Vogt, and J. Arkko, "End-Host Mobility and Multihoming with the Host Identity Protocol", Internet Society, RFC 5206, April 2008. Available from: *http://tools.ietf.org/html/rfc5206*

[23] D. Eastlake and P. Jones, "US Secure Hash Algorithm 1 (SHA1)," Internet Society, RFC 3174, September 2001. Available from: http://tools.ietf.org/html/rfc3174

[24] R. Rivest, "The MD5 Message-Digest Algorithm," Internet Society, RFC 1321, April 1992. Available from: http://tools.ietf.org/html/rfc1321

[25] B. Xu, Y. Liu, X. He, and Y. Tao, "On the Architecture and Address Mapping Mechanism of IoT," 2010 International Conference on Intelligent Systems and Knowledge Engineering (ISKE), pp. 678-682, November 2010.

[26] Pappas, S. Hailes, and R. Giaffreda, "Mobile Host Location Tracking through DNS, " London Communications Symposium (LCS) and Photonics London, LCS 2002, Available from: *http://www.ee.ucl.ac.uk/lcs/previous/LCS2002/lcs2002.html*

[27] T. Aura, "Cryptographically Generated Addresses (CGA)," Internet Society, RFC 3972, March 2005. Available from: *http://tools.ietf.org/html/rfc3972*

[28] EM Microelectronics, "EM4100 / Read Only Contactless Identification Device," data sheet, 2004. Available from *http://www.datasheetarchive.com/*

[29] EM Microelectronics, "EM4450/4550 / 1 KBit Read/Write Contactless Identification Device," data sheet, 2010. Available from: *http://www.datasheetarchive.com/*

[30] D. Johnson, C. Perkins, and J. Arkko, "Mobility Support in IPv6," Internet Society, RFC 3775, June 2004. Available from: *http://tools.ietf.org/html/rfc3775*

Chapter 8

Advancements and Prospects of Forward Directional Antennas for Compact Handheld RFID Readers

Ahmed Toaha Mobashsher and
Rabah W. Aldhaheri

Additional information is available at the end of the chapter

1. Introduction

In the current age of rapid technological progress and development, radio frequency identification (RFID) technology has found many applications in various areas such as supply chain, warehouse, and retail store management. Alike mobile communications, the high data performance and compact profile are becoming obvious expectations of the users of handheld RFID devices. In this regard, directional antennas have a bright prospect for a more user friendly experience even in the rugged environments.

This chapter presents a comprehensive review of RFID technology concerning the prospects of directional especially forward directional antennas and propagation for multi-band operation. The technical considerations of directional antenna parameters are also discussed in details in order to provide a complete realization of the parameters in pragmatic approach to the directional antenna designing process, which primarily includes scattering parameters and radiation characteristics. The antenna literature is also critically overviewed to identify the possible solutions of the directional antennas to utilize in single and multi-band handheld RFID reader operation. However, it has been seen that these techniques can be combined to enhance the directional antennas with wider bandwidths and higher gain. Last but not least, the possibilities of forward-directional antennas which spectacularly use the surface wave for the radiation will be explored, and the difference with the conventional directional antennas with them will be discussed.

2. Prospects of forward directional antennas for handheld RFID readers

Radio frequency identification (RFID) technology having a huge potential with higher production efficiency, real-time inventory updates, greater product security and restricting counterfeit products, is becoming rapidly engaged in production, transportation, and retailing of products. On the investment front, more than 40% of shippers have increased their investment in RFID technology for supply chain applications (Research and Markets 2012). RFID technology is rapidly evolving due to (Cole P. H. 2003):

- increased awareness of the technology;

- development of improved techniques for multiple tag reading;

- realization in the business community of the benefits of widespread adoption in the supply chain;

- adoption by designers of sensible concepts in the arrangement of data between labels and databases;

- development of efficient data-handling methodologies in the relevant supporting communication networks;

- appreciation of the need for cost reduction, and

- development of new manufacturing techniques that will achieve manufacture of billions of labels at acceptable costs.

The global RFID market is projected to reach US$18.7 billion by the year 2017 where growth will be primarily driven by fast paced deployments of RFID projects in developing Asian countries, especially in China. Developments in the field of smart labels are projected to hold the key to future revenue growth (Research and Markets 2012). However, RFID is now at a stage where there are potentially large benefits from wider application but still some barriers remain.

It is widely known that handheld RFID readers need more compactness in design than fixed or mounted readers. Thus it is more challenging to make the designs more compact to meet the expectations of the users. Similar to mobile communications, the multi-standard capability, high data performance and compact profile are becoming obvious expectations of the users of RFID devices. Among the frequency bands that have been assigned to RFID applications, higher-frequencies have the advantage of high data transfer rate with far field detection capability (Islam et al. 2010).

Directional antennas usually radiates in a directive manner. They force the electromagnetic energy into a specific and desired direction. This type of antenna decreases the interferences of other tags in the undesired direction, while also increasing the reading range as well, since the gain of the directional antennas are higher than the omni-directional antennas.

In order to reduce the overall size of the handheld RFID readers, the need to reduce the size of the antenna is highly essential. But reducing the size of antenna limits its performances.

Also when the operating frequency of RFID systems rises to the microwave region (2.45/5.8 GHz bands), the reader antenna design becomes more delicate and critical. This is especially true when a directional antenna is needed for handheld applications. However, it is a popular practice for handheld RFID readers to assemble a vertically radiating directional antenna in right angle with the reader; thus the radiation literally becomes front-directional to the reader (Fig. 1). This arrangement significantly increases the actual RFID reader profile. Hence it is greatly advantageous for a compact RFID reader to produce antennas with front-directional radiation patterns.

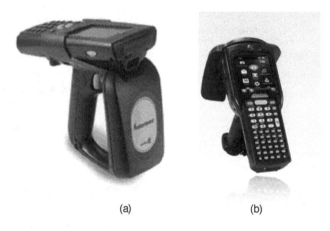

(a)   (b)

Figure 1. Handheld RFID reader with an external antenna module (a) (Ukkonen et al. 2007), (b) (MC3190-Z Handheld RFID Reader 2012).

3. Evolution of forward directional antennas & limitations: A literature scenario

In this section we will focus on the development of multi-band antenna designing process. In last few years, there are a lot of antennas multi-band antennas has been designed, but most of them are omni-directional in radiation manner. But still some novel structures can be found in recent researches, that provide multi-band operation with directional radiation characteristics (Li et al. 2012, Mobashsher et al. 2010, Sabran et al. 2011). However, all of these antennas use a big metallic ground plane in order to reflect the radiation from the patch. Hence the actual applied antenna profile is bigger than the patch alone. So these antennas are not suitable for portable RFID applications. Another technique is widely applied in antenna domain in order to achieve directional radiation patters- the utilization of surface waves (also called trapper waves) (Zucker 1993).

In literature, surface-wave or end-fire antennas are mostly used to produce front-directional radiation patterns. Folded dipole (Fan et al. 2009) and folded (Yang et al. 2010) antennas are reported to have this type of radiation. Although these antennas produce good directional patterns, they are inappropriate for compact multi-band applications as they are printed in both sides of the substrate and resonate only in one operating frequency. The reported conformal (Dong & Huang 2011) and plate (Yao et al. 2011) end-fire antenna with good radiation patterns have a bulky profile and are unsuitable for a portable use. It is worth to mention that for handheld compact applications, uni-planar antennas are more beneficial than double-sided microstrip in terms of compactness and the integration capability with solid-state active and passive components. The uni-planar compact yagi antenna, reported in (Nikitin & Rao 2010), is difficult to be incorporated with the circuitry due to its construction.

Several quasi-Yagi (Kan et al. 2007) and bow-tie (Eldek et al. 2005) antenna provides wide bandwidth, but they do not give flexibility to choose specific frequencies of operation, thus in turn increases interference with neighboring operating bands. The frequency reconfigurable planar quasi-Yagi antennas (Qin et al. 2010) are also unsuitable for its complex feeding structure. A printed dipole (He et al. 2008) with etched rectangle apertures on surface has reported to have dual-band characteristics; but it suffers mostly in the consistency of the radiation patterns. Again, these are mostly double sided planar antennas. A multi-band Quasi-Yagi-type antenna is reported in (Ding et al. 2011). However, the feeding transition takes a wide area which in turn increases the antenna size significantly. It is obvious that the front directionality of the antennas will provide the handheld RFID readers a compacter solution and there is a huge research interest in this area in recent years to achieve optimum solution for the practical application.

4. Possibilities of forward-directional antennas for compact handheld RFID readers: An example from single & multi-band perspectives

Forward directional antennas can provide the RFID readers the desired compactness both in single and multi-band applications. In this section an example of forward directional antenna is discussed which is enhanced from single band to multi-band operation. The frequency of operation is chosen from the microwave ISM bands (2.45/5.8GHz), while the same methodology and design procedures are applicable for any combination of narrow RFID bands as the design has much flexibility. The details of the design process and the optimization are also discussed for the better understanding of the readers.

The flow chart in Fig. 2 describes the design and fabrication process of a desired single antenna. Firstly the requirements of the RFID antenna are collected according to the specifications of the RFID reader module. An extensive literature is reviewed for the design purpose. In this case some models are chosen which provides proper forward directionality. These are discussed in previous section. However, every design has some advantages and also some disadvantages. Proper understanding of the antenna operation is effective for proper chose of the antenna model. Meanwhile the familiarization with the simulation software IE3D EM simulator has been performed and an appropriate model is chosen for the design by using mathematical models. The design is simulated and its performance is examined and optimized until satisfactory result is obtained. When the optimized antenna is achieved, it is time to enhance the performances by using some techniques. This section is discussed more in the next sections. The last but not the least step is the fabrication process where the prototype is going to be built and finally the measurement of the antenna parameters for validation and comparison with the simulated results. At the end of this process the desired antenna with the given specification is attained. However, if any step fails to achieve its objectives, it is repeated again until the aims are met.

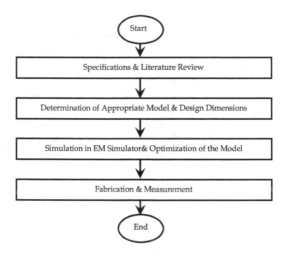

Figure 2. Flow chart design and fabrication process of a desired single band RFID antenna

The dual-band operation of the any antenna is an additional advantage. The design procedure of the dual band antenna is slightly different from the previously discussed the single band antenna. The design process is illustrated in the Fig. 3.

In order to meet the requirements of dual-band RFID reader antenna specifications, first the lowest cutoff frequency of the operation should be met. It is because of the fact that at the lowest frequency means the biggest wavelength in comparison to other higher frequencies and hence the respective current path should be the longest. Thus, the basic dimension of the antenna is defined by the lowest operating frequency. Electromagnetic simulation is an advanced technology to yield high accuracy analysis and design of complicated microwave and RF printed circuit, antennas and other electronic components. In antenna designing process, after the determination of the antenna dimensions with suitable model, the next step is to simulate the design in suitable electromagnetic software. The modeling and formulation are mainly derived through the use of Green's functions.

In simulation there may be some disagreements with calculated designed dimensions. In that case, the antenna should be optimized varying the parameters. When the lowest frequency is attained, then as the next step, further simulation and optimization is needed to meet the next operational frequency band with the same antenna. The possible solution is to achieve multi-resonance by employing notches, slots or additional current paths. Similarly, when optimizing, it should be taken care that the lowest frequency should not shift from the desired frequency. At the end of this design process, the antenna goes for prototyping and performance evaluation process.

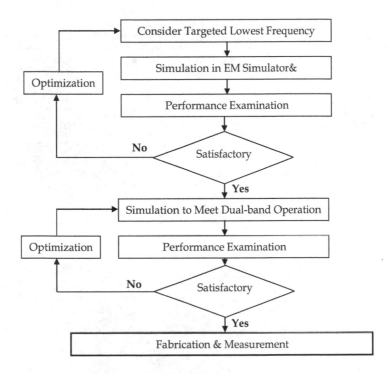

Figure 3. Flow chart simulation process of the multi-band RFID antenna

4.1. Antenna geometry

The schematic diagram of the proposed antenna is exhibited in Fig. 4. The antenna is fabricated on a low-loss substrate of medium permittivity (Rogers TMM4 ε_r = 4.5, $tan\delta$ = 0.002) with height 1.52 mm. Metallization of 1 oz copper cladding is used in only one side, which makes the antenna uni-planar and suitable to incorporate into the circuitry of the RFID reader. Also, the fabrication process of the antenna is relatively easy and cost effective. The antenna consists of a microstrip feeding line, two unsymmetrical ground planes and a folded strip with a small top branch. The width of the CPW feed line is fixed at F_w = 3mm. In order to achieve 50Ω characteristic impedance, the feeding line section (X-axis) is separated by a gap of g = 0.3mm from both right and left sides of ground plane. At the end, the antenna is fed by a 50Ω SMA coaxial connector from the side. Three triangular periodic open-end stub (POES) cells are infixed on the upper edge of each side of ground plane. The POESs are symmetrical with respect to the center line of the feeding strip in longitudinal direction (Y-axis). The POESs are optimized to improve the impedance matching of the antenna with no other effect on other characteristics of the antenna.

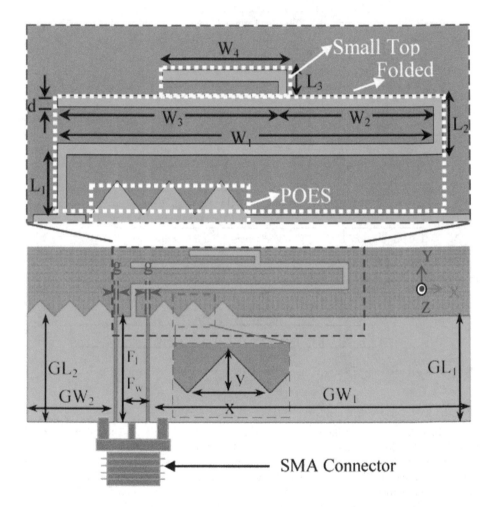

Figure 4. Geometric structure of the proposed antenna

4.2. Design procedure

Fig. 5 shows the design procedure of the proposed antenna. In design process, the lower band was first designed, since the antenna profile is usually circumscribed by the wavelength of lower frequencies. Inspired from (Yang et al. 2010) Design A was optimized to operate in the lower band with a small and uni-planer orientation. The 3D full-wave commercial package, Ansoft HFSSv10 was utilized in assisting the optimization of the antenna.

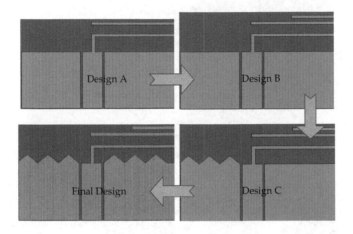

Figure 5. Adopted design steps of the proposed antenna

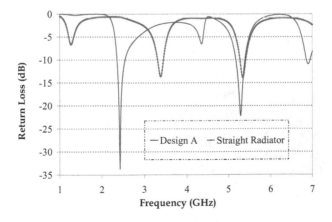

Figure 6. Comparison between straight and folded design (Design A)

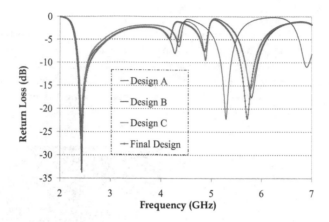

Figure 7. Improvement in impedance matching

Figure 8. Surface current distributions of the antenna at 2.45 GHz

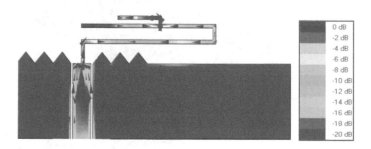

Figure 9. Surface current distributions of the antenna at 5.8 GHz

As shown in Fig. 6, it was observed that folded configuration is better suited for a good impedance matching and directional characteristics; while the radiating strip is placed straight, it acts as a long monopole antenna, resonating (f_{r1}) in some lower value. When folded, the coupling effects among the horizontal parts and ground plane appear and the radiating strip acts like a quasi-folded dipole antenna.

Design B was then introduced to provide desired dual band characteristics. The basic folded strip was designed to support the lower frequency band of ISM 2.45GHz and the forth resonance (f_{r4}) was dragged down to the desired 5.8GHz by introducing the small top branch. Thus small top branch confirms the proper selection of upper band to ISM 5.8GHz. However, it is the main folded strip which supports the small top branch for its radiation and impedance matching. Next, three optimized POES cells were inserted on the top edge of the left co-planar ground plane in Design C, which provides better impedance matching. Lastly, the final optimal design (Final Design) of the proposed antenna was derived by introducing another three POES cells in the upper left edge of the right ground plane. This orientation of POES cells improves both the impedance matching and bandwidths of both the desired bands (f_{r1}, f_{r4}). It is noted that the second resonance (f_{r2}) is generated mostly from the lower arm of basic folded strip; and the third resonance (f_{r3}) is influenced by the upper arm. Hence introduction of the small top branch vitally changed f_{r3}, while the resonance did not change with the application of POES. On the other hand the matching of f_{r2} varies by the affixation of POES cells. Fig. 7 describes the improvement in impedance matching through the design procedure.

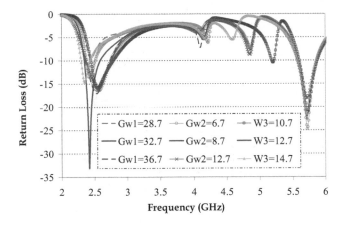

Figure 10. Comparison of Return Loss of the proposed antenna as a function of G_{w1}, G_{w2} & W_3 which dominate only lower operating band

In order to gain further understanding of the way resonances are excited, we also examine surface current distributions of the proposed antenna extracted from the full-wave, method-of-moment based electromagnetic simulator Zeland IE3D. From Fig. 8 & 9, it is evident that at both resonating frequencies, the current density is indeed higher on the folded strip of the antenna, thus the dimensions of the folded strip are assumed to govern both bands. The top small branch is active only in the upper frequencies, and has so effect for the 2.45 GHz resonance. So the optimal value of top small branch is vital for 5.8 GHz operation. Nevertheless, it is noted that the ground plane do not resonate in any of the desired resonances, but it provides better impedance matching for the desired bands. It is seen that the triangular POES cells increase the current path in the ground plane without influencing the currents on the radiating strip; hence only effects the scattering parameters, not radiation characteristics.

4.3. Optimization, parametric analysis & guidelines

A parametric analysis of the proposed antenna was carried out in order to illustrate the optimization process of the proposed antenna. The results are exhibited in Figs. 11 to 14. All the parameters have been studied to find the impact of the impedance matching, especially on the resonance frequencies and bandwidth.

It is observed that both the bands varied in terms of resonance, whenever the dimensions of the lower portion of folded strip, like L_1, L_2, W_1 and W_2 are changed. The length, x of the triangular POES cell do not have much effect on the antenna performances, but the height, y is very crucial for the bandwidth of the lower band and adjusting the upper band. Also, the number of POES cells are important for adjusting the impedance matching of both bands.

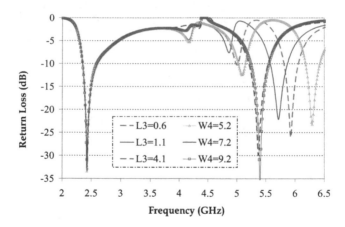

Figure 11. Performance comparison of Return Loss of the proposed antenna for the variation of L_3 & W_4 which dominate only higher operating band

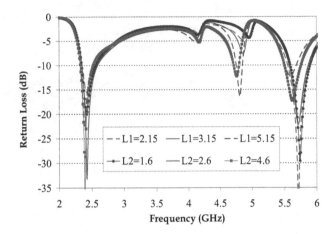

Figure 12. Return Loss comparison of the proposed antenna as a function of L_1 & L_2 which dominate both the higher and lower operating bands

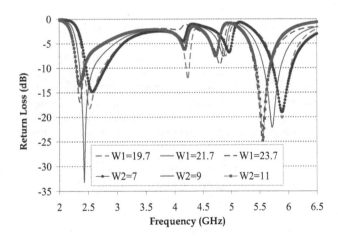

Figure 13. Comparison of Return Loss Vs Frequency of the proposed antenna when the values of W_1 & W_2 are varied that dominate both the operating bands

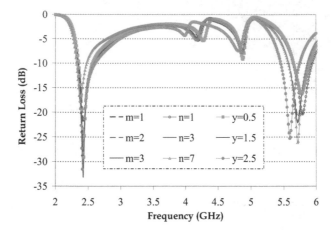

Figure 14. Dependencies of lower & higher operating bands for various heights (y) of the POES as well as for various left POES numbers (m) & right POES numbers (n)

The bandwidth and resonating frequency of the lower frequency band can be further improved by increasing the structure of the ground plane. However, in that case the return loss degrades and the antenna becomes bigger in size. As a design guideline, it is suggested that the dimension of W_3 is very useful for tuning up the lower resonance without changing the overall antenna profile. It is also feasible to generate another resonance near 2.45GHz by adding another radiating upper element with the existing one. But the high electromagnetic coupling makes it difficult to match the reflection coefficient; even in some cases the antenna might loose its front-directive radiation characteristics when achieving multi resonance near lower operating band.

After dealing with the lower band, the upper band can be tuned by proper selection of L_3 and W_4. Nevertheless, if the investigated front-directional antenna was required to operate in a number of discrete bands for higher frequencies, a set of top branches equal to the number of the resonating bands could be used. In that case, the coupling effect should be carefully eliminated by adjusting the width, W_4 and distance, L_3 for each top branch.

These are illustrated in Table 1 for better realization of the antenna geometry. The calculated parametric values of the radiating strip are based on the guided quarter wavelength of the substrate of the predicted dominating portion at the resonances; the rest are assumed in an arbitrary manner. Afterwards all the parameters are optimized through empirical observations.

Parameters	Calculated Value	Optimized Value	Variation	Lower Band			Upper Band		
				f_{r1}	$\|RL\|$	BW	f_{r4}	$\|RL\|$	BW
GW_2	-	8.7	>	↑	↓	↑	-	↑	-
			<	-	↓	↓	-	↓	-
GW_1	-	32.7	>	↑	↓	↑	-	-	-
			<	-	↓	↓	-	-	-
$GL_1=GL_2$	-	10	>	↓	↓	↓	-	↓	-
			<	↑	↓	↓	~	-	~
L_1	-	3.15	>	↓	↓	-	↓	↓	↓
			<	-	↓	-	-	↑	↑
L_2	-	2.6	>	↓	↑	-	↓	↓	-
			<	-	↓	-	↑	↑	-
L_3	-	1.7	>	-	-	-	↓	↑	-
			<	-	-	-	↑	↑	-
W_1	19.7	21.7	>	↓	↓	↓	↓	↑	-
			<	↑	↓	↑	↑	↓	-
W_2	12.9	9	>	↓	↓	↓	↓	↑	-
			<	↑	↓	↑	↑	↓	-
W_3	12.9	12.7	>	↓	↓	↓	-	-	-
			<	↑	↓	↑	-	-	-
W_4	6.8	7.2	>	-	-	-	↓	↑	-
			<	-	-	-	↑	~	-
x	-	3	>	-	-	↓	-	-	-
			<	-	-	-	-	-	-
y	-	1.5	>	-	-	↑	↓	↑	-
			<	-	-	↓	↑	↓	↓
m	-	3	>	-	↑	↑	-	↑	↑
			<	-	↓	↓	-	↓	↓
n	-	3	>	↓	↓	↓	~	~	-
			<	-	↓	↓	↑	↓	↓

* '-' represents the frequency phenomenon is independent for the increment of the parameter. '~' represents the non-monotonic fluctuation of the criteria upon increasing the geometry. '↑' and '↓' represent the enhanced and deteriorated phenomenon of the antenna upon changing parameter-values.

Table 1. Sensitivity of the antenna resonance frequencies (f_{r1}, f_{r4}), return loss ($|RL|$) and bandwidth (BW) when varying geometric parameters

4.4. Prototyping & measurements

The antennas were fabricated with the optimized parameters for experimental verification. A photograph of the prototypes is presented in Fig. 15. An Agilent N5230A PNA-L network analyzer was used to measure the electrical performance of the prototype. The simulated and measured return loss of the prototype is presented in Fig. 16. A good agreement between the simulated and measured results is observed. The small difference between the measured and simulated results is due to the effect of SMA connector soldering and fabrication tolerance. The measured return loss curve shows that the proposed antenna is excited at 2.45 GHz band with a –10 dB return-loss bandwidth of 320 MHz (2.35–2.67 GHz) and at 5.8 GHz band with an impedance bandwidth of 310MHz (5.6–5.91 GHz). The maximum return loss of -28.4dB and -34.2dB is obtained at the resonant frequencies of 2.46 GHz and 5.76 GHz respectively. Most of the desired frequencies are below -15dB level. The narrowband characteristics are useful to minimize the potential interferences between the RFID system and other systems using neighboring frequency bands such as UWB, WiMAX etc.

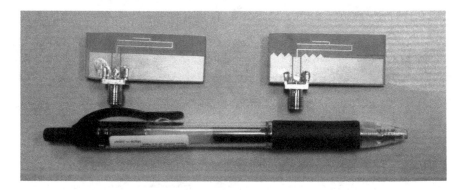

Figure 15. Photograph of the fabricated prototypes

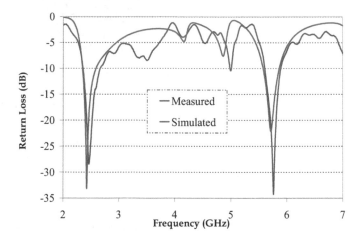

Figure 16. Return loss comparison of the proposed prototype

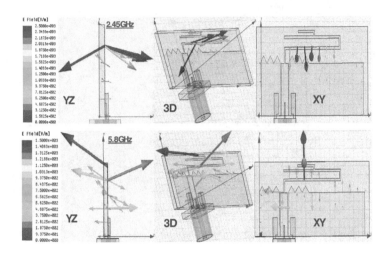

Figure 17. Electric field vector distribution of the proposed antenna

The electrical field vector distribution of the proposed antenna at frequencies 2.45 and 5.8 GHz is illustrated in Fig. 17. This distribution is extracted from HFSS software. It is noticed that the vital electric fields are generated from the folded resonating portion. The middle horizontal portion of the strip generates the radiation and the upper horizontal directs the energy propagation towards the end fire direction. The ground plane acts more likely as a reflector or suppressor. It mainly suppresses the back radiation and improves the impedance matching of the radiation element. The electric fields generated from the top edge of the ground plane are observed to extend in the forward direction. Thus it forces the electromagnetic energy and produces the front-directional radiation patterns. The measured radiation patterns of the fabricated prototype antenna at 2.45 and 5.8 GHz are illustrated in Fig 18. It is seen that the antenna provides front-directional radiation pattern for both bands. More importantly the cross-polarization levels are low (at least -10 dB) in both E- and H-planes. Also the front to back ratio in the scale of -10 dB is observed in the lower resonance; and at upper band it increases around the scale of -20 dB. The peak gain of the prototype is found to be 3 and 3.2 dBi at 2.45 and 5.8 GHz respectively.

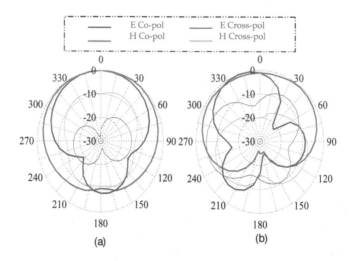

Figure 18. Measured normalized radiation patterns of the fabricated prototype at (a) 2.45 and (b) 5.8 GHz

4.5. Guidelines of future works

The design procedure of the presented dual-band antenna is applicable for multi-band extension. Following this design and optimization steps a more efficient antenna covering triple band including UHF and ISM microwaves is quite feasible. However, the antenna profile will increase when the operating frequency decreases, but yet the front directionality of the antenna is very effective for the compact handheld RFID reader design.

5. Conclusion

This chapter reveals the advantages and limitations of forward directional antennas to the readers for compact handheld RFID operation for multi-band operation. A comprehensive review and limitations of RFID technology concerning the prospects of directional antennas and propagation for both single and multi-band operation are presented in this chapter. The technical considerations of directional antenna parameters are also discussed in details in order to provide a complete realization of the parameters in pragmatic approach to the directional antenna designing process, which primarily includes scattering parameters and radiation characteristics. The antenna literature is also critically overviewed to identify the possible solutions of the directional antennas to utilize in single and multi-band handheld RFID reader operation.

Author details

Ahmed Toaha Mobashsher and Rabah W. Aldhaheri

Department of Electrical and Computer Engineering, Faculty of Engineering, King Abdulaziz University, Jeddah, Saudi Arabia

References

[1] Cole P. H. 2003. Fundamentals in Radiofrequency Identification. Summer Course, MIT, http://autoidlab.eleceng.adelaide.edu.au/Papers/RFID_World_Fundamentals.pdf

[2] Ding, Y., Jiao, Y.C., Fei, P., Li, B. & Zhang, Q. T. 2011. Design of a Multiband Quasi-Yagi-Type Antenna With CPW-to-CPS Transition. IEEE Antennas and Wireless Propag. Lett., 10:1120-1123.

[3] Dong, L. & Huang, K.-M. 2011. A novel conformal end-fire antenna design using the competitive algorithm of simulating natural tree growth. Prog. Electromagn. Res. C, 24:207-219.

[4] Eldek, A.A., Elsherbeni, A.Z., & Smith, C.E. 2005. Wide-Band Modified Printed Bow-Tie Antenna With Single and Dual Polarization for C– and X-Band Applications. IEEE Trans. Antennas Propag., 55(9):3067-3072.

[5] Fan, Z.G., Qiao, S., Huangfu, J.T. & Ran, L.X. 2009. A miniaturized printed dipole antenna with V-shaped ground for 2.45 GHz RFID readers. Prog. Electromagn. Res. Lett., 6:47–54.

[6] He, Q.-Q., Wang, B.-Z. & He, J. 2008. Wideband and Dual-Band Design of a Printed Dipole Antenna. IEEE Antennas and Wireless Propag. Lett., 7:1-4.

[7] Islam, M. T., Mobashsher A. T., & Misran N. 2010. Design of a microstrip patch antenna using a novel U-shaped feeding strip with unequal arm. Electronics Letters. 46(14):968-970

[8] Kan, H. K., Waterhouse, R. B., Abbosh, A. M. & Bialkowski, M. E. 2007. Simple Broadband Planar CPW-Fed Quasi-Yagi Antenna. IEEE Antennas and Wireless Propag. Lett., 6:18-20.

[9] Li, R. L., Cui, Y. H., & Tentzeris, M.M. 2012. Analysis and Design of a Compact Dual-Band Directional Antenna. IEEE Antennas and Wireless Propagation Letters. 11:547-550.

[10] MC3190-Z Handheld RFID Reader, 2012. http://www.motorola.com/Business/US-EN/Business+Product+and+Services/Mobile+Computers/Handheld+Computers/MC3190-Z+Handheld+RFID+Reader

[11] Mobashsher, A. T., Islam, M. T., & Misran, N. 2010. A novel high gain dual band antenna for RFID reader application. IEEE Antennas and Wireless Propag. Lett., 9: 653-656.

[12] Nikitin, P. V. & Rao K. V. S. 2010. Compact Yagi Antenna for Handheld UHF RFID Reader. IEEE Antennas and Propag. Soc. Int. Symp., pp 1-4.

[13] Qin, P.-Y., Weily, A.R., Guo, Y.J., Bird, T.S. & Liang, C.-H. 2010. Frequency Reconfigurable Quasi-Yagi Folded Dipole Antenna. IEEE Trans. Antennas Propag., 58(8): 2742-2747.

[14] Research and Markets, 2012 "Radio Frequency Identification (RFID) Technology - Global Strategic Business Report", http://www.researchandmarkets.com/research/ 63wt75/radio_frequency_id. 20 August, 2012.

[15] Sabran, M. I., Rahim, S. K. A., Rahman, A. Y. A., Rahman, T. A., Nor, M. Z. M., & Evizal. 2011. A dual-band diamond-shaped antenna for RFID application. IEEE Antennas Wireless Propag. Lett., 10: 979–982.

[16] Ukkonen, L., SydÃnheimo, L. & Kivikoski, M. 2007. Read range performance comparison of compact reader antennas for a handheld UHF RFID reader [supplement, applications & practice]. IEEE Commun. Mag., 45:24–31.

[17] Yang, X., Yin, Y.Z., Hu, W. & Zhao, G. 2010. Compact printed double-folded inverted-L antenna for long-range RFID handheld reader. Electron. Lett., 46(17):1179-1181.

[18] Yao, G., Xue, Z., Li, W., Ren, W. & Cao, J. 2011. Research on a new kind of high directivity end-fire antenna array. Prog. Electromagn. Res. B, 33:135-151.

[19] Zucker, F. J. 1993. Surface-wave antennas. Antenna Engineering Handbook, 3rd Edition, Richard C. Johnson, Edited, McGraw-Hill Inc.

Chapter 9

Design and Implementation of RFID-Based Object Locators

T. S. Chou and J. W. S. Liu

Additional information is available at the end of the chapter

1. Introduction

In the coming decades, an increasingly larger number of baby boomers will grow into old age. This trend has led to an increasing demand for devices and services (e.g., [1-8]) that can help elderly individuals to live well and independently. *Object locator* is such a device. The device can assist its users in finding misplaced household and personal objects in a home or office. Figure 1 shows several object locators offered today by specialty stores and websites. Each of these locators contains an interrogator with a few buttons and an equal number of tags: Even the largest one, the leftmost one in the figure, offers only 8 buttons. The buttons are of different colors, and there is a tag of the color matching the color of each button. By attaching a tag to an object to be tracked, the user can look for the object by pressing the button of matching color on the interrogator. The tag attached to the object beeps and flashes in response and thus enables the user to find the object. Other locators work similarly.

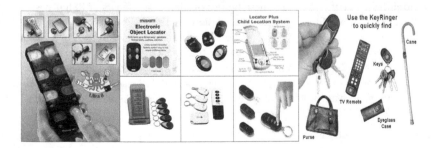

Figure 1. Existing object locators

Existing object locators are not ideal in many aspects: The number of buttons on the interrogator and tags is fixed, and the number is small. Extending the locator to track more objects is impossible. – If the user were to use more than one tag of the same color, the tags would all respond to the search signal for tag(s) of the color from the interrogator. This situation is clearly not desirable. – When a tag breaks, the user must purchase a replacement tag of the same color as the broken one. Tags are battery-powered. A tag might become a lost object itself after it runs out of battery. More seriously, the interrogator itself can be misplaced. Obviously, these are serious shortcomings.

This chapter describes three designs and a proof-of-concept prototype of object locators based on the *RFID (Radio Frequency Identification)* technology. RFID-based object locators do not have the drawbacks of existing object locators. In particular, RFID-based object locators are extensible, reusable, and low maintenance. They are extensible in the sense that the maximum number of tracked objects is practically unlimited and that a RFID-based object locator can support multiple interrogators. The interrogator software can run on a variety of platforms (e.g. desktop PC, PDA, smart phone and so on). A mobile interrogator can be tagged and thus, can be searched via other interrogators when it is misplaced. Reusability results from the fact that all RFID tags used for object locators can have globally unique ids. Hence, tags never conflict, and a tag can be used in more than one object locators. Low maintenance is one of the advantages of RFID technology. One of the designs uses only RFID tags without batteries; the user is never burdened by the concern that a tag may be out of battery.

This chapter makes two contributions: The first is the object locator designs presented here. The designs use different hardware components and have different hardware-dependent software requirements. The information provided by the chapter on these aspects should enable a developer to build a suitable object locator platform, or an extension to one of the commonly used computer and smart mobile device platforms. The functionality of hardware-independent object locator software is well defined, and a C-like pseudo code description can be found in [9].

The hardware capabilities and object search schemes used by the designs lead to differences in search time and energy consumption. We provide here a numeric model that can be used to determine the tradeoffs between these figures of merit. Developers of RFID-based object locators can use the results of the analysis as design guides. Today, object locators based on all designs are too costly: Typical RFID readers have capabilities not needed by our application and cost far more than what is suitable for the application. Through this analysis, we identify the design that is the most practical for the current state of RFID technology and project the advances in the technology required to make RFID-based object locators affordable (i.e., with prices comparable with some of the locators one can now find in stores.) This is the second contribution of the chapter.

The rest of this chapter is organized as follows. Section 2 describes closely related works. Section 3 describes use scenarios that illustrate how a RFID-based object locator may be used. Section 4 presents three designs of RFID-based object locators. Section 5 describes the implementation of a proof-of-concept prototype based on one of the designs. It also describes the reader collision problem [10] encountered in the prototype and the solution we use to deal with the problem. Section 6 describes a numeric model for computing energy consumption and search time and compares the merits of the designs. Section 7 concludes the chapter and discusses future works.

2. Background and related work

This section first presents a brief overview of RFID technology as a way to state the assumptions made in subsequent chapters on state-of-the-art readers and tags. Our object locator resembles location detection systems in its goal: assisting users to locate objects. The section describes existing location systems and compare and contrast them with our object locators.

2.1. RFID technology

RFID technology is now applied to a wide spectrum of applications. As an example, personal identification application is used to provide authentication and authorization to individuals carrying their RFID tags so that they can be automatically identified by a central computer. Card-like RFID tags used as smart cards in public transports is another example: Information on money stored in a tag is automatically deducted when the card holder presents the card in front of a reader while getting on or off a transporter. Other applications include using RFID tags as markings of books for more efficient library management, shipping containers for tracking them by retail industry, and so on.

Figure 2 shows a typical system that uses RFID technology. The host machine uses one or more RFID readers to retrieve digital information stored in RFID tags and processes the information according to the needs of one or more applications. In general, a RFID tag contains a globally unique identification (UID) as well as data fields organized in a standard way [11]. A RFID-based object locator only needs the UID information; other data fields are not used.

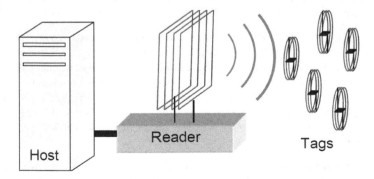

Figure 2. A configuration of RFID system

There are three types of RFID tags: passive, semi-passive and active. A *passive tag* has no internal power source: It gets the power it needs to operate from the incident RF signal radiated by a reader. The readable distance of such a tag ranges from 10 cm to a few meters depending on the frequency of the incident RF signal and its antenna design. In contrast, *semi-passive* and *active tags* have internal power source. Semi-passive tags can increase their readable distances by leveraging internal power. Like passive tags, semi-passive tags respond only after receiving some command from the reader. An active tag, on the other hand, can send RF signals to a reader even when it is not commanded by the reader. Being battery free and having long lifetime (in tens of years) are the major advantages of passive tags over other types of tags for our application.

Each message sent from a reader to tags contains a command code. Among the sets of commands defined by ISO15693 [12], our object locators use only *mandatory* commands and *custom* commands. Standard-compliant tags support all commands in the mandatory set. Commands in the custom set are defined by tag IC manufacturer according to application needs.

The command used to read UID of a tag is the *inventory* command in the mandatory set. This command has only the non-addressed mode, while the other commands have both non-addressed and addressed modes. A command in the *non-addressed mode* is processed by all tags which receive it. A command in the *addressed mode* consists of the command code followed by a UID. When a tag receives an addressed-mode command, it first checks whether the UID is its own. The tag processes and responds to the command only when it is the tag addressed by the UID.

2.2. Location detection systems

Many different location detection systems are available today. Global Positioning System (GPS) [13] is the most well known. Priced at about $ 100 US each, GPS navigators are widely used in cars, buses and so on. However GPS has its limitations. Reflection, occlusion and multipath effects seriously interfere with distance measurement and make GPS ineffective indoors. For this reason, indoor location detection systems use a variety of other technologies.

Active Badge [14] is representative of infrared-based location detection systems. A badge containing an infrared transmitter is attached to each object to be tracked by the system. The transmitter sends periodically messages containing the unique identification of the badge. The messages are caught by some infrared receivers at fixed known locations and relayed by the receivers to a central computer. The central computer resolves the position of the badge based on the locations of the receivers. Shortcomings of systems such as Active Badge arise from the fact that infrared signals cannot penetrate most materials in a building and are easily interfered by other infrared sources.

Ultrasound is used to assist with distance measurement in Bat [15] and Cricket [16]. These systems use both ultrasound and RF signals to measure distances between beacons (transmitters) and listeners (receivers): When a beacon at a known location transmits an ultrasound signal and a RF signal concurrently, a listener can calculate the distance to the beacon from the difference between the arrival times of the signals.

Many indoor location detection systems use RF-based technology to take advantage of the fact that RF signals penetrate most non-metallic materials. RADAR [17] is an example. The system estimates distance by estimating the strength of RF signals. Specifically, the system measures in the initialization phase at a set of fixed locations the strengths of a RF signal sent by a location-known transmitter. The measured strengths are stored in a database to be used later as yardsticks during the working phase. In the working phase, each receiver measures the strength of a RF signal transmitted from a tracked object and sends the strength to a central computer. The computer compares the measured strengths with the information stored in the database and then resolves the possible position of the transmitter (i.e., the tracked object). MoteTrack [18], similar to RADAR, uses empirical distance measurement to estimate positions of objects. WLAN (wireless local area network) can be used to build location detection systems also. SpotON [19] and Nibble [20] are examples.

Compared with the above mentioned location detection systems, an object locator must be a far more low cost solution and must be ultra easy to set up and use. Many indoor location detection systems (e.g. Bat and Active Badge) rely on a big infrastructure or a pre-computed database (e.g. RADAR) to support location estimation. These systems are too costly to deploy and maintain and hence, unsuitable for home use. Cricket system provides a low cost location-aware service. An object with a receiver can determine its location. This is not what an object locator does. A misplaced object does not need to know its own location; the user looking for it needs to know.

3. User scenarios

The routine usage of an object locator requires only three operations: Add, Delete and Query. We describe these operations here to illustrate how a locator may be used. Without loss of generality, we assume that a new object locator kit contains a portable interrogator, a dozen of RFID tags and agents. As illustrated by Figure 3, the interrogator resembles a smart phone. It has a small non-volatile storage and a RF transceiver together with a network address. We will return in the next section to describe how the RF transceiver is used, as well as what agents are and do. Unlike common smart phones, however, the interrogator has a RFID reader. The reader is used for the Add operation described below.

Specifically, Figure 3 shows parts of the user interface on an interrogator with a LCD touch screen and two buttons. The LCD touch screen is used as both input and output user interface. A user can select an item among the items displayed on the screen, the button at the bottom left corner to confirm a selection, and the button at the bottom right corner to cancel the selection. Some operations need text input. The virtual keyboard shown on right is for this purpose.

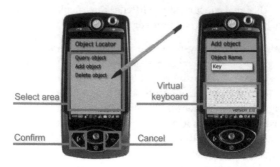

Figure 3. Object locator user interface

Figures 4 and 5 illustrate Add and Query operations, respectively. Add operation works in a similar way as the address book of a smart phone. Using this operation, the user can add the registration of an object to be tracked into the interrogator. By registration, we mean a mapping between the id of the tag attached to an object and the name of the object. The user queries the locations of objects by their names. In response to a query, the interrogator uses the object-name-tag-id mappings to resolve which one of the registered objects to search. Figure 4 shows a scenario: The user picks an unused tag and attaches it to an object to be tracked as shown in Figure 4(a) and (b). Then, the user puts the tag close to the interrogator and selects Add object. This step is shown in Figure 4(c). In response to Add object command, the interrogator reads the id of the tag, displays a new text field and prompts the user to enter a name (e.g., Key). When the user confirms the name, the interrogator creates a mapping associating the name with the id of the tag attached to the object, and stores the mapping in its local non-volatile memory. This is illustrated in Figure 4(d). The user repeats the above steps to register each object until all objects to be tracked are registered.

Figure 4. Add operation

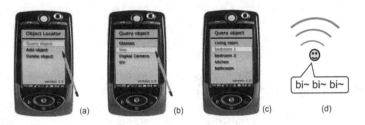

Figure 5. Query operation

Query operation is the work horse of the object locator. The user presses Query object on the touch screen, as illustrated by Figure 5(a), to invoke this operation for assistance in finding misplaced objects. When the names of registered objects are displayed, the user selects the object to be searched; in this example, it is Key. After the user confirms the selection, as shown in Figure 5(b), the interrogator retrieves from its local storage the id of the tag attached to the object with the selected name and starts a search for the tag with that id. Hereafter, we call the tag being searched the *queried tag* and the object attached to the tag the *queried object*. We will describe the search process in the next section.

Object locators of different designs present the result of Query operation in different ways. As examples, Figure 5(c) and (d) shows two different responses. In Figure 5(c), the interrogator directs the user to the place (e.g. bedroom 1) where the queried object is found. In Figure 5(d), the queried tag beeps, allowing the user to look for it by following the sound. This version works like the existing locator described in Section I.

Delete operation removes the registration of an object, i.e., the object-name-tag-id mapping stored in the interrogator: The user can invoke the operation by pressing Delete object on the touch screen. In response, the interrogator displays the list of registered objects, allowing the user to select the object (e.g. Key) to be deleted. The interrogator deletes the mapping after the user confirms the selection. Delete operation frees the tag attached to the now unregistered object and makes the tag free for use to track some other object.

4. Alternative designs

The three designs of object locator are called Room-level Agents, Interrogator and Tags (RAIT) locator, Desk-level Agents, Interrogator and Tags (DAIT) locator and Desk-level and Room-level Agents, Interrogator and Tags (DRAIT) locator. As their names imply, each of the locator consists of tags, agents and at least one interrogator. The adjectives room-level and desk-level describe the ranges of RFID readers used by the designs. The ranges of room-level readers and desk-level readers are sufficiently large to cover a typical-size room or desk, respectively.

The term tag refers specifically to RFID tags. Each tag has a unique id, hereafter called *TID*. One of the designs uses only passive tags. The other designs call for tags that can beep upon receiving query messages containing their TIDs. It is possible to implement such tags using semi-passive RFID tags since the battery in such a tag can be used not only to improve read range but also to drive a beeper.

An *agent* is a device that aids the interrogator in locating the queried object (i.e., the queried tag). Each agent has a RF transceiver, together with a programmable network address, a RFID reader, and a RFID tag. The RFID reader in the agent enables the agent to search for the tags within its coverage area. As stated in Section III, the interrogator also has a RF transceiver with a network address. This allows the interrogator and all agents to form a wireless local area network (WLAN). The network address of the interrogator (or each interrogator in a multiple-interrogator system) is unique and so is the network address of each agent. The interrogator requests assistance from an agent by sending the TID of the queried tag to the agent via the WLAN. We assume that the network provides reliable communication. We do not mention other aspects of the WLAN because they are not relevant to our discussion.

4.1. RAIT locator

A disadvantage of the existing locator is that a user needs to walk around the house when searching an object and the interrogator needs to repeatedly send the query signal until the user hears the queried tag or gives up the search. RAIT locator is designed to eliminate this disadvantage.

RAIT locator uses one or more agents to cover each room, and the house is fully covered by agents as shown in Figure 6. When the user invokes a Query operation, the interrogator sends a query message containing the TID of the queried tag to agents and thus requests the agents to search the queried tag on its behalf. Each agent broadcasts an addressed mode read request with the TID retrieved from the query message to read the tags within range. The tag with id matching the TID beeps upon receiving a read request, in addition to responding to the agent. The agent finding the queried tag reports its network address to the interrogator. This information enables the interrogator to display the results illustrated by Figure 5(c), telling the user to go to the specified room where the queried object has been found.

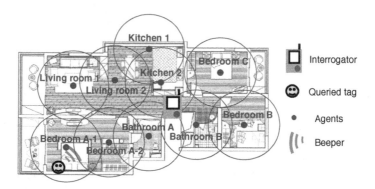

Figure 6. Configuration of RAIT locator

Obviously, the agents must be set up before a RAIT locator can be used. Figure 7 lists the steps carried out by the user and work done by the system during the set up process. The goal of Steps 3-5 is to make sure that there is no blind region. A *blind region* is an area where tags cannot be read by any agent. The corners of a room are the most likely to be blind regions. This is the rationale behind Step 3. When the *TEST READ RANGE* switch of an agent is on, the agent repeatedly broadcasts non-address mode read messages. In this way, the agent enables the user to determine whether any of the corners is a blind region in Step 3.

---

1. Choose a location near middle of a room and temporarily attach an agent to the ceiling or furniture at the location.
2. Turn on *TEST READ RANGE* switch on the agent.
3. Pick up a tag and check whether the tag beeps at each corner of the room.
4. If no, adjust the location of the agent or add one more agent at another location in the room and turn on *TEST READ RANGE* switch on the additional agent. Then go back to Step 3. If yes, turn off *TEST READ RANGE* switch.
5. Securely attach the agents tested in Steps 2-4 at their respective locations.
6. Put the interrogator near the agent and execute *Register Agent operation*.
7. Repeat step 1 to 6 until all agents covering the house are registered.

Figure 7. Agent set-up process

The Register Agent operation in Step 6 is similar to Add operation described in Section 3. Its goal is to assign a human-readable location name to an agent, so that the interrogator can later generate query results illustrated by the example in Figure 5(c). During the operation, the interrogator prompts the user to provide a unique name for the location of each agent. For example, if the living room needs two agents, Living Room R(ight) and Living Room L(eft) are good names for them.

The interrogator also assigns a unique network address to the agent being registered. The id of the tag in an agent is the product serial number of the agent. The interrogator uses the id to distinguish the agent from previously registered agents. By assigning successive network addresses to agents as they are registered and initialized one by one, successive Register Agent operations enable each initialized agent to join the WLAN and later compute the addresses of other agents by adding or substituting some number from its own address.

Figure 8 depicts the format of messages in a RAIT locator. This format supports multiple interrogators: The src_addr allows agents to identify the interrogator issuing the query message. The dest_addr allows them to address their responses to a specified interrogator. Data field allows interrogators to synchronize their databases created by Add and Register Agent operations. We will discuss how the other fields are used shortly.

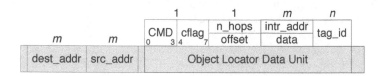

Figure 8. RAIT locator message format

4.2. DAIT and DRAIT locators

DAIT locator, shown in Figure 9, is an extension of RAIT locator. The designs are similar in how the Query operation is handled by the interrogator and agents. DAIT differs from RAIT primarily in the required read ranges of agents. The read range of agents used in a DAIT locator is less than one meter. Agents with such a small range offer higher accuracy in locations of queried tags. Information on the agent that finds the queried tag tells the user the location of the searched object within a small vicinity of the agent. Tags in DAIT locators are passive; they do not beep because a user can easily find the misplaced object even though the tag does not beep. Because tags do not need to beep, they can be battery free. This is a major advantage of DAIT locator.

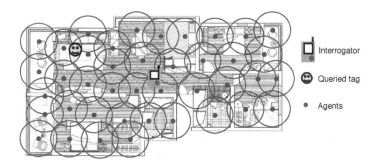

Figure 9. Configuration of DAIT locator

However, it is significantly more complicated to set up desk-level agents. Blind regions of RAIT locator are easy to detect and eliminate because a blind region is typically created by walls and is near the read boundary of an agent. In the case of DAIT locator, a room cannot be fully covered by one or two agents. Any three adjacent agents may create a blind region.

Our solution is to give a user a circular thread whose circumference is less than $3\sqrt{3}$ (i.e., the circumference of a regular triangle whose center is one unit away from its corners) times their read range and instruct a user to set any three adjacent agents within the circular thread. By doing so, blind regions never occur.

DRAIT locator has a hybrid design that aims to extend the lifetime of semi-passive tags. A DRAIT locator contains both room-level and desk-level agents. Its interrogator asks desk-level agents to search first. The interrogator asks room-level agents only when no desk-level agent finds the queried object. We set up desk-level agents on furniture in addition to setting up room-level agents as described above. Because misplaced objects are often on furniture or in vicinities of them, the queried object can often be found by a desk-level agent, and the tag on it does not need to beep.

4.3. Search schemes

A queried object can be searched in three ways: broadcast, relay and polling. The *broadcast scheme* is the most straightforward. The interrogator broadcasts a query message with the tag_id field filled with TID of the queried tag. The agents finding the queried tag report their agent ids to the interrogator and the others do not reply.

The knowledge on the agent network addresses and the number of agents enables an interrogator to request assistance from agents one at a time using the *relay scheme*: To search for a queried tag, the interrogator sends a query message containing its own address in intr_addr field, the number of agents to be queried in n_hops and the TID of the queried tag in tag_id to the first agent: The simplest choice is the agent with the smallest address. In response to a query, each agent searches for the tag with the TID in its own cover area. The agent reports its own address to the interrogator if it finds the tag; otherwise it decreases n_hops by one, increments its own network address by one to get the address of the next agent and then forwards the query message to the next agent.

According to the *polling scheme*, the interrogator also sends a query message to the first agent in its polling list, provides the agent with the TID of the queried tag and waits for response from the agent. The agent replies to the interrogator no matter whether it finds the tag or not. If the response from an agent is negative, the interrogator sends the query message to the next agent in its polling list. Advantage of the polling scheme over the relay scheme is that the interrogator can dynamically alter the search sequence.

5. Prototype implementation

We implemented a proof-of-concept prototype of DAIT locator, the design that does not require customized semi-passive tags. Indeed, all components used in our prototype are readily available today. Parts (a) and (b) of Figure 10 show an agent and the portable interrogator of our prototype, respectively. The agent is composed of a microcontroller, a RF transmitter, a RF receiver and a RFID reader module. The microcontroller is ATMEL ATmega128. It runs at 8MHz and has 128k bytes flash / 4k bytes EPPROM. The RF transmitters and receivers interconnecting interrogator(s) and agents are LINX TXM(RXM)-433-LR, which use 433MHz ASK. RFID reader modules are MELEXIS EVB90121, which is ISO15693-compliant and uses a directional antenna. We use TI OMAP5912 and NEC Q-VGA to implement the portable interrogator. The current version of our prototype supports the three operations described in Section II and uses the polling search scheme.

The lack of customized antenna design for tags and readers and the reader collision problem seriously affects the performance of our prototype. Our DAIT prototype uses only tags with directional antennae. (Again, the reason is that such tags are readily available.) When the antennae of tags and readers are directional, the read performance of agents depends on the orientation of the antennae. Clearly, tagged objects may be placed in arbitrary orientations. As a consequence, it is impossible to ensure optimal or near optimal alignment of the tag antennae towards the agents covering their locations. This is the reason that tags in a DAIT object locator should have omni-directional antennae. Agents with omni-directional antennae can be simply set on furniture as shown in Figure 11(a). Agents with directional antennae should be attached to the ceiling as shown in Figure 11(b). This arrangement requires a read range of 2-3 meters. With readers of a sufficiently large read range, RAIT locators can use tags with directional antennae without performance concern.

(a)   (b)

Figure 10. Agent and interrogator

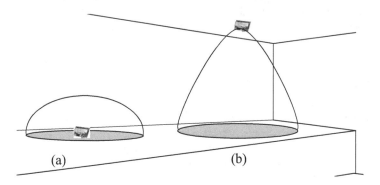

Figure 11. Arrangement of agents

Close proximity of readers (i.e., agents) is necessary in order to avoid blind regions. Our DAIT prototype is no exception. When RFID readers have overlap coverage areas, signals sent at the same time from them to tags in the overlap region interfere with each other. This is called the *reader collision problem* [10]. Fortunately, only the broadcast scheme suffers this problem. Our prototype uses the polling scheme to avoid the problem: According to the polling schemes (or the relay scheme), agents search the queried tag in sequence; signals from readers never interfere.

A DAIT locator that uses the broadcast scheme can circumvent the reader collision problem in many ways. For example, the DAIT prototype can let each agent delay transmitting its query signal by an amount of time that is a function of its network address. In this way, agents try to avoid transmitting query signals at the same time. This solution is practical and easy to implement.

Another solution requires each agent to know the network addresses of its neighbors. Each agent can be viewed as a node in a connected graph. There is an edge between two nodes when the agents represented by them have overlapping coverage regions. A graph coloring algorithm can be used to assign different colors to adjacent nodes. The reader collision problem never occurs as long as agents labeled by different colors do not transmit query signals concurrently. This solution is likely to have a better response time than the solution mentioned above or the relay and polling schemes. However it requires additional hardware for each agent to automatically detect its neighbors or connectivity information entered by the user manually. The additional hardware makes agents more costly, and complicated operations by the user make an object locator hard to use.

6. Relative merits

We use search time and energy consumption of a single query to measure the relative merits of object locator designs. Search time and energy consumption per query depend on many factors including the number of agents, search scheme, search sequence and locations of misplaced objects.

6.1. Search time and energy consumption

The expressions of energy consumption and search time per query according to broadcast, relay and polling schemes are listed in Table 1. The expressions assume that agents and interrogator(s) are battery powered and communicate in the manners described in Section 4. The notations used in the expression are defined in Table 2.

The total energy consumed by the object locator for processing a Query operation according to the broadcast scheme is the sum of the three terms in the first row of Table 1. In this case, the interrogator transmits only one query message per Query operation. The energy it consumes is E_{IA}. The energy consumed by each agent in the search is E_{Arfid}. The total energy consumed by all agents is $N_A(x, y, r)E_{Arfid}$, where $N_A(x, y, r)$ is the number of agents with range r in a rectangular space of dimensions x and y. The agent finding the queried tag consumes E_{AI} to send a response back to the interrogator.

In the expressions, pA_i denotes the probability that the i-th agent in the search sequence finds the queried tag. In general, this probability is a function of the number and location distribution of objects (i.e., tags) in the house. (To keep the expressions simple, our notations do not show this dependency.)

broadcast	E_{total}	$E_{IA} + N_A(x, y, r)E_{Arfid} + E_{AI}$
	T_{avg}	$D_{IA} + pA_1(D_{Arfid} + D_{AI}) +$ $\sum_{i=2}^{n}(\prod_{k=1}^{i-1}1 - pA_k)pA_i(iD_{Arfid} + D_{AI})$
relay	E_{avg}	$E_{IA} + pA_1(E_{Arfid} + E_{AI}) +$ $\sum_{i=2}^{n}(\prod_{k=1}^{i-1}1 - pA_k)pA_i(iE_{Arfid} + (i-1)E_{AA} + E_{AI})$
	T_{avg}	$D_{IA} + pA_1(D_{Arfid} + D_{AI}) +$ $\sum_{i=2}^{n}(\prod_{k=1}^{i-1}1 - pA_k)pA_i(iD_{Arfid} + (i-1)D_{AA} + D_{AI})$
polling	E_{avg}	$E_{IA} + pA_1(E_{Arfid} + E_{AI}) +$ $\sum_{i=2}^{n}(\prod_{k=1}^{i-1}1 - pA_k)pA_i(iE_{Arfid} + (i-1)E_{IA} + iE_{AI})$
	T_{avg}	$D_{IA} + pA_1(D_{Arfid} + D_{AI}) +$ $\sum_{i=2}^{n}(\prod_{k=1}^{i-1}1 - pA_k)pA_i(iD_{Arfid} + (i-1)D_{IA} + iD_{AI})$

Table 1. Expressions for search time and energy consumption

- D_{IA}: Delay of a message transmitted from an interrogator to an agent
- E_{IA}: Energy consumption of a message transmission from an interrogator to an agent
- D_{AA}: Delay of a message transmitted from one agent to another
- E_{AA}: Energy consumption of a message transmission from one agent to another
- D_{AI}: Delay of a message transmitted from an agent to an interrogator
- E_{AI}: Energy consumption of a message transmission from an agent to an interrogator
- D_{Arfid}: Time for an agent to use its RFID reader to search a queried tag
- E_{Arfid}: Energy consumption of a RFID reader in an agent per search of a queried tag

Table 2. Notations

The expression of the expected time taken by the locator using the broadcast scheme to respond to a Query operation assumes that agents search the queried tag in sequence in order to avoid the reader collision problem. The first term in the expression is the time taken by the query message from the interrogator to reach all the agents. If the first agent finds the queried tag, which occurs with probability pA_1, the addition delay is $D_{Arfid} + D_{AI}$. This is the reason for the second term in the expression of T_{avg}. In general, the probability that the queried tag is found by the i-th agent is $\prod_{k=1}^{i-1}(1-pA_k)pA_i$. When this occurs, each of the other agents spends D_{Arfid} amount of time to search for the queried tag before the i-th agent can respond to the interrogator. Hence, the delay is $iD_{Arfid} + D_{AI}$.

The average search time of an object locator that uses the relay and polling scheme are estimated by the expressions in the fourth and sixth rows in Table 1, respectively. Relay and polling scheme also lets all agents search the queried tag in sequence. This is why the coefficients in these expressions are the same as the coefficients in the expression of T_{avg} for the broadcast scheme. The expressions of the average energy consumption can be derived from the expressions of the average search time by substituting energy consumption for message transmission delay because sending a message cause both transmission delay and energy consumption.

As stated earlier, Table 1 is based on the assumption that agents and the interrogator are battery powered. Hence, the total energy consumption includes energy consumptions of agents and an interrogator. However, agents can be connected to wall plugs, especially when the number of agents is small, as in the case of RAIT locators. The interrogator using relay and broadcast scheme consumes exactly E_{IA} to search a queried tag. The interrogator using polling scheme consumes at least E_{IA} to search a queried tag. Therefore, the polling scheme is suitable for stationary interrogator(s) and the relay and broadcast scheme are suitable for portable interrogator(s) if we do not need to account for the energy consumption of agents.

6.2. Model of object locality

The probability pA_i of that an agent A_i finds the queried tag, and hence the misplaced object, depends on where the object is at the time. To calculate this probability, we use a locality model of tracked objects. The model gives the spatial probability density of the locations of each object. For the sake of simplicity and without noticeable lose of accuracy, we partitions the space in the search area into unit squares, rather than treating the coordinates of a location as continuous variables. (Except for where it is stated otherwise, the dimension of a unit square is 1 cm by 1 cm.) This allows us to model a house as a finite, discrete and planar search space. We denote the space by $Z = \{Z_{x,y}\} \subseteq N \times N$. Each element $Z_{x,y}$ of the space is a unit square; its location is given by the coordinate (x, y) where both x and y are integers. All agents are at fixed and known locations. A misplaced object may be placed anywhere within the search space.

We call the probability of finding a queried object at $Z_{x,y}$ the (*existence*) *probability* of the object at $Z_{x,y}$. (For example, if we find an object at $Z_{x,y}$ on the average 10 times in 100 searches for the object, the (existence) probability of the object at $Z_{x,y}$ is approximately 0.10. We use $pZ_{x,y}(j)$ to denote the existence probability of an object with a tag of id $= j$ at $Z_{x,y}$. We do not consider the situation where someone has taken some registered object shopping, for example, while someone else is searching for it in the house. Hence, for every object being searched, the sum of the probabilities of it being at all locations in the search space equals to 1.

Figure 12 gives an illustrative example. The figure is not drawn scale, and each unit square in this example is 10 cm by 10 cm in dimension. Two agents A_1 and A_2 are at their locations. The id of A_1 is 1 and the id of A_2 is 2. The rectangle models a desk. It contains 15 unit squares. The number in each square gives the probability of a queried object being at the location. Since the numbers add up to 1, they tell us that the object is surely somewhere in the rectangle. We want to calculate pA_i, the probability that the agent with id $= i$ can find the queried tag. Using Figure 12 as an illustrative example, we see that pA_1 equals to the sum of all existence probabilities within the read range of the agent A_1; in other words, pA_1 is about 0.87. Similarly, we find that pA_2 is about 0.68.

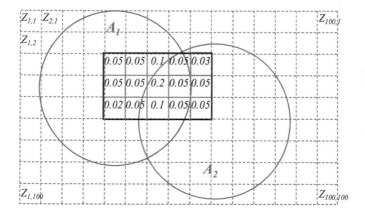

Figure 12. Locality model

We call the area where a misplaced object might be placed an *object region*. The size of an object region is the total area of the region in number of unit squares. We characterize the locality of a misplaced object by the size and shape of its object region and its existence probabilities of being at each unit square within the region. Once we know the locality parameters of an object and coverage area of each agent A_i, the terms pA_i can easily be calculated. We can then calculate the average search time and energy consumption of the object based on the probability pA_i for all agents.

6.3. Evaluation environment and results

The environment we used to evaluate the relative performance of our designs has a 10m by 10m search space, containing 1000 × 1000 unit squares of size 1 cm by 1 cm. Agents are placed according to the arrangement in Figure 13(a). The number of agents is $N_A(1000, 1000, r)$. Again, r is the read range of an agent. The ranges of desk-level and room-level agents are 100 and 350, respectively, the typical number of room-level agents in a RAIT locator is $N_A(1000,1000,350) = 6$, and the typical number of agents in a DAIT locator is equal to $N_A(1000,1000,100) = 42$.

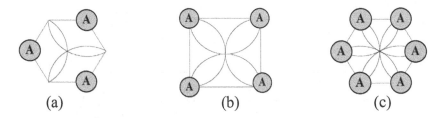

Figure 13. Possible arrangement of agents

In Section 4, we said that the agent with the smallest network address is the first agent and the other agents are asked one by one in order of agent ids to search for the queried object. We call this search order *sequential*. Alternatively, we can ask the agents in non-increasing order of their empirical existence probabilities. This search sequence is called *profiling*.

Our evaluation program assumes that object regions are circular for the sake of simplicity. The center and radius of an object region are randomly generated. The variables D_{IA}, D_{AA} and D_{AI} in Table 2 have the same values because both interrogators and agents use the same kind of RF transceiver. For the same reason, E_{IA}, E_{AA} and E_{AI} have the same value. For convenience, we use D_{Arfid} and E_{Arfid} as base units of delay and energy consumption. The ratio of D_{IA}/D_{Arfid} (D_{AI}/D_{Arfid} and D_{AA}/D_{Arfid}) is called *DRatio* and the ratio of E_{IA}/E_{Arfid} is called *ERatio*. The evaluation program needs only these two parameters rather than all variables.

Figure 14(a) and (b) show the average search time for broadcast scheme, relay scheme, and polling scheme (i.e., polling in sequential order), as well as polling scheme with profiling. The search time of relay and polling schemes is higher than broadcast scheme for all values of DRatio. The search time of polling scheme with profiling is less than that of broadcast scheme when DRatio is less than about 1 (10^0) for $N_A = 42$ and 1.25 ($10^{0.1}$) for $N_A = 6$.

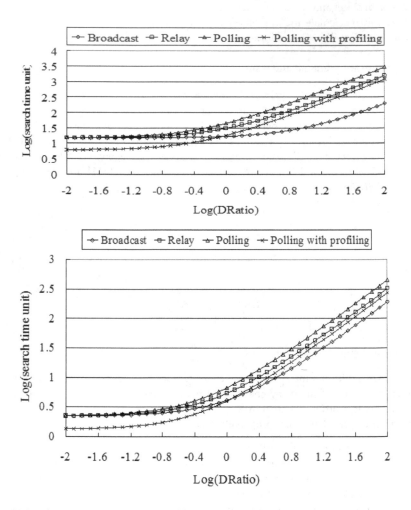

Figure 14. Search time Vs DRatio: (a) top NA = 42; (b) bottom NA = 6

Figure 15 shows the average energy consumption consumed by agents when N_A is 42 and 6. The energy consumption consumed by agents is the same, when the relay and polling scheme is used. As Figure 15(a) depicts, the energy consumption of relay scheme and polling scheme are the same. Their consumptions and that of polling scheme with profiling is less than that of broadcast scheme when ERatio is less than 1.99 ($10^{0.3}$) and 7.94 ($10^{0.9}$), respectively. Values of ERatio at the intersections of the curves in Figure 15(b) are about 3.16 ($10^{0.5}$) and 15.85($10^{1.2}$).

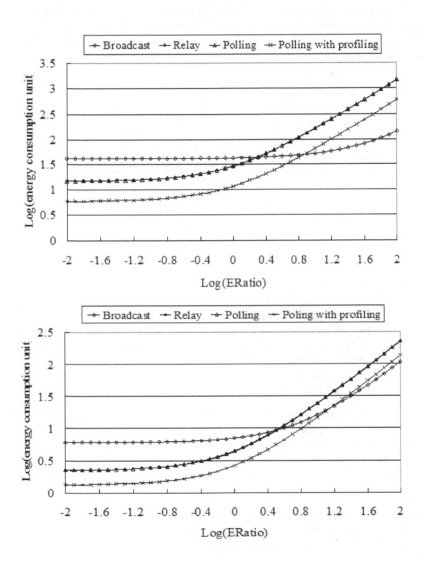

Figure 15. Energy consumption of agents Vs ERatio: (a) top NA = 42; (b) bottom NA = 6

Table 3 gives a summary. The table suggests the broadcast scheme when DRatio is high and search time is more important than energy consumption. When DRatio is low, the differences among the search times of all search schemes are small. Energy consumption becomes the dominant factor for comparison. It is possible for agents in a RAIT locator to connect to power source. For energy saving on interrogators, we suggest polling scheme with profiling for stationary interrogators and relay or broadcast scheme for portable interrogators. As for DAIT locators, we consider energy consumption of an interrogator and agents. We suggest polling scheme with profiling when ERatio is low and the same suggestions as that for a RAIT locator if ERatio is high.

	Low DRatio Low ERatio	Low DRatio High ERatio	High DRatio Low ERatio	High DRatio High ERatio
RAIT locator (fewer agents)	PI: *broadcast* or *relay* SI: *polling with profiling*	PI: *broadcast* or *relay* SI: *polling with profiling*	*broadcast*	*broadcast*
DAIT locator (more agents)	*polling with profiling*	PI: *broadcast* or *relay* SI: *polling with profiling*	*broadcast*	*broadcast*

PI: portable interrogator   SI: stationary interrogator

PI: portable interrogator; SI: stationary interrogator

Table 3. Summary of suggested search schemes

7. Conclusion

We described here three alternative designs for RFID-based object locator. These object locators are extensible, reusable and low maintenance. They are easy for users to set up and use. Our analysis shows that search time and energy consumption for all designs and search schemes depend the capabilities of RFID readers and RF transceivers used by agents. Roughly speaking, polling and relay schemes are competitive to broadcast scheme only when DRatio or ERatio are less than 10.

We implemented a proof-of-concept DAIT prototype object locator to demonstrate the object locator concept and designs. The prototype uses only readily available hardware components, including readers and tags with directional antennae. The performance of the prototype is far from ideal, primarily for this reason. Because it is impossible to control the orientation of tag antennae, omni-directional antennae are better suited for our application.

The total cost of an object locator depends on many factors. The total hardware cost of a minimum object locator is the sum of the costs of an interrogator and required number of agents and tags. Compared with the costs of interrogator and agent, the hardware cost of tags is significantly lower and, for the discussion here, can be neglected.

Currently, the total hardware cost of an object locator is dominated by the total cost of agents, and the cost of an agent is dominated by the RFID reader in the agent. The number of agents required to fully cover a house depends on dimensions x and y of the house, the read range of the agents and the way agents are placed. To get a rough estimate, we assume that the coverage area of each agent is a circle. Figure 13 depicts three ways to place agents. Putting agents further apart than locations shown in Figure 13(a) can create blind regions. Putting more agents closer than those indicated in Figure 13(c) is not necessary since the space is covered by at least two agents. We need six room-level agents to cover a 10m x 10m space even when we place agents as shown in Figure 13(a) (i.e., as far as possible without creating blind regions). The existing object locator costs $ 50 US. A RAIT locator is not competitive to the existing locator unless the cost per room-level agent is about $ 10 US. As for DAIT locator, the cost per desk-level agent must be much lower. We are optimistic that the cost of agents will become sufficiently lower in the coming decade as the need for more and more products (e.g., Smart pantry [1], dispenser in [4]) containing RFID readers are developed to take advantage of this technology.

Acknowledgment

This work is partially supported by the Taiwan Academia Sinica thematic project SISARL (Sensor Information System for Active Retirees and Assisted Living (http://www.sisarl.org).

Author details

T. S. Chou[1] and J. W. S. Liu[2]

1 Computer Science Department, University of California at Irvine, Irvine, CA, USA

2 Institute of Information Science, Academia Sinica, Nangang, Taipei, Taiwan

References

[1] Hsu CF., Liao HY., Hsiu PC., Lin YS., Shih CS., Kuo TW., Liu JWS., Smart Pantries for Homes: Proceedings of IEEE International Conference on Systems, Man and Cybernetics, SMC2006, 8-11 October 2006, Taipei, Taiwan; 2006. p4276-4283.

[2] Yeh HC., Hsiu PC., Tsai PH., Shih CS. and Liu JWS. APAMAT: A Prescription Algebra for Medication Authoring Tool: Proceedings of IEEE International Conference on Systems, Man and Cybernetics, SMC2006, 8-11 October 2006, Taipei, Taiwan; 2006. p3676-3681.

[3] Hsu YT., Hsiao SF., Chiang CE, Chien YH., Tseng HW., Pang AC., Kuo TW., Chiang KH. Walker's Buddy: An Ultrasonic Dangerous Terrain Detection System: Proceed-

ings of IEEE International Conference on Systems, Man and Cybernetics, SMC2006, 8-11 October 2006, Taipei, Taiwan; 2006. p4292 – 4296.

[4]   Tsai PH., Yu CY., Shih CS., Liu JWS. Smart Medication Dispensers: Design, Architecture and Implementation. IEEE Systems Journal 2011; 5(1). p99-110.

[5]   Medication Reminders: http://www.epill.com/ (accessed 01 July 2012)

[6]   Senior Care Products: http://www.agingcare.com/Products (accessed 01 July 2012)

[7]   Alababa.com, showroom for object locators: http://www.alibaba.com/showroom/object-locator.html (accessed 01 July 2012)

[8]   Hsu CC., Chen JH. A novel sensor assisted and RFID-based indoor tracking system for elderly living alone: Sensor 2011; 11. p10094-10113.

[9]   Chou TS. Design and Implementation of Object Locator. MS thesis, Taiwan National Tsing Hua University, 2006

[10] Leong Kin Seong LK., Ng ML., Cole PH. The Reader Collision Problem in RFID Systems: Proceedings of IEEE International Symposium on Microwave, Antenna, Propagation and EMC Technologies for Wireless Communications, MAPE 2005, August 2005. p 658 – 661.

[11] EPCglobal standards overview: http://www.gs1.org/gsmp/kc/epcglobal (accessed 11 July 2012)

[12] ISO15693 Standard, OpenPCD: http://www.openpcd.org/ISO15693 (accessed 12 July 2012)

[13] Getting I. The Global Positioning System: IEEE Spectrum 1993; 30(12). p36–47.

[14] Want R., Hopper, A., Falcao V., Gibbons J. The Active Badge Location System: ACM Transactions on Information Systems. 1992 10(1). p91–102.

[15] Harter A., Hopper A. A New Location Technique for the Active Office. IEEE Personal Communications; 1997 4(5). p43–47.

[16] Nissanka BP., Chakraborty A., Balakrishnan A. The Cricket Location-Support System: Proceedings of ACM International Conference on Mobile Computing and Networking (MoBiCOM), 6 – 11 August, 2000, Boston MA USA. 2000. p32-43.

[17] Bahl P., Padmanabhan V. RADAR: An In-Building RF-based User Location and Tracking System: Proceedings of IEEE The Conference on Computer Communications (INFOCOM), 26 - 30 March, 2000, Tel-Aviv, Israel. 2000. p775-784.

[18] Konrad L., Welsh M. MoteTrack: A Robust, Decentralized Approach to RF-Based Location Tracking: Proceedings of the International Workshop on Location and Context-Awareness at Pervasive, 12 – 13 May 2005, Oberpfaffenhofen near Munich, Germany. 2005. p489-503.

[19] Hightower J., Borriello G., Want R. SpotON: An Indoor 3-D Location Sensing Technology Based on RF Signal Strength. University of Washington, Seattle, Tech. Rep. 2000-02-02, February 2000.

[20] Castro P., Chiu P., Rremenek T., Muntz RR. A Probabilistic Room Location Service for Wireless Networked Environments: Proceedings of ACM International Conference on Ubiquitous Computing (UBICOMP), 30 September – 2 October 2, 2001, Atlanta, Georgia USA; 2001. p18-34.

Chapter 10

RFID Textile Antenna and Its Development

Lukas Vojtech, Robi Dahal, Demet Mercan and
Marek Neruda

Additional information is available at the end of the chapter

1. Introduction

Textile fabric material has become one of the most important things in life. In early times people used to wear the animal skin to cover their body. The advance form of this is all the clothes we wear today. They protect our body from changing environment conditions and keep us warm. As the technology is increasing day by day, it is influencing every sector. With the increase in wireless technology the electromagnetic radiation also increases. This increased radiation may affect human body severely. Thus with invent of problem, cure was also proposed to make a conductive textile material that could be equally wearable but at the same time work as a filter and does not allow the harmful frequency signal to penetrate into the human body. This completely changed the purpose of fabric material which was previously assumed to be used only for keeping human body warm as now it can be used for protection against the harmful electromagnetic radiation.

Going one step further ahead, we have tried to explore more advantage of textile fabric. With this new invention of conductive textile, we have designed an antenna for RFID (Radio Frequency Identification) applications made out of conductive textile material.

2. RFID basics

The RFID uses wireless technology to identify the objects. It consists of RFID tag and a reader. The bi directional communication between the tag and the reader is accomplished by the Radio Frequency (RF) part of the electromagnetic spectrum, to carry information between an RFID tag and reader. There are two types of RFID tag. Passive RFID tags are the ones that does not require any external power supply and works by receiving the signal from reader and retransmit the signal back to reader. Active RFID tag consists of external source in them. These are more complex than passive RFID tags and also give long range communication between tag and reader, when compared with passive tag.

The basic block diagram describes the bi directional communication between the tag and the reader, see Figure 1. The tag antenna in the block diagram receives the RF signal from the reader. This signal is received by the tag antenna, rectified and supplied to the chip to power it up. After the chip is powered up, it now acts as a source and retransmits the signal back to the reader. The reader after receiving the signal sends further to the computer to process the data. The method used to send the signal back to the reader from the tag is called back scattering.

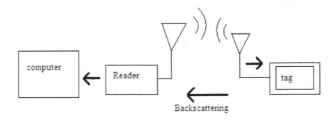

Figure 1. Block diagram of RFID system

2.1. RFID transponder

RFID Transponder is basically a radio transmitter and receiver. It mainly consists of two parts, antenna and the integrated circuit (IC). The main function of an antenna is to capture the radiated electromagnetic field by the reader at a definite frequency. The received electromagnetic energy is converted to electrical power and supplied to integrated circuit. The IC chip in the transponder has the capability to store the information to be transmitted to the reader, execute the series of command and also sometimes stores new information sent by the reader [1]. The IC chip mainly consists of a rectifier which rectifies the alternating voltage (AC) received by antenna to the continuous voltage (DC) and supplies to the rest of the circuit in the IC chip.

The IC used for the research is EM4222. This is a read only UHF identification device. The EM4222 is used as a passive chip for UHF transponder. It does not have any internal power supply source. The RF beam is transmitted by the reader. The antenna in the transponder receives the signal, rectifies it and supply the rectified voltage to the chip. The basic block diagram is shown in Figure 2.

From the block diagram, it can be seen that the radio beam is received at the terminal A in the chip. This signal is rectified to a DC voltage. The shunt regulator is used to limit the input voltage to the logic circuit. It also protects the Schottky diode which is used as a rectifier.

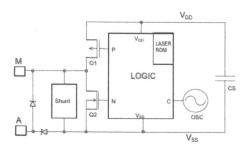

Figure 2. Block diagram UHF transponder EM4222 [2]

The on chip oscillator in the transponder is used to provide the clock pulse to the logic and also defines the data rate. On chip oscillator present in the transponder oscillates at a frequency of 512 kHz.

If the supply voltage is less than the threshold voltage, the oscillator and the logic cannot function properly and thus the transponder cannot be activate. At this condition, the logic is in reset position. This ensures that the transistor Q2 is off during power up and do not let any false operation to act.

Among the two transistors, Q1 is turned on during power up. Q2 is the modulation transistor which when turned on, loads the antenna with the information from the tag. Q2 is active when the data is to be transmitted from transponder to reader.

In order to have a maximum power transferred from antenna to the chip, the antenna should be designed such that the impedance of the antenna is conjugated matched with that of chip for the given frequency. Generally the chip has capacitive impedance so to have a perfect match the antenna impedance should be inductive in nature.

Parameter	Symbol	Test conditions	Min	Type	Max	Units
Oscillator frequency	Fosc	-40°C to +85°C		512		KHz
Wake up voltage	Vwu	VM-VA rising		1.4		V
Static Current Consumption	I STAT	VM=1v	400	1	600	µA
Input Series Impedance	Zin	869 MHz ; -10dBm	1.0	128-j577	1.8	Ω
Input Series Impedance	Zin	915 MHz ; -10dBm		132-j553	5	Ω
Input Series Impedance	Zin	2.45 GHz ; -10dBm		80-j232		Ω

Table 1. Electrical characteristics of IC EM4222, VM-VA=2V, TA=25°C, unless otherwise specified [2]

2.2. RFID matching

The tag antenna receives RF energy from the reader. The tag antenna works for a definite resonant frequency. So when the reader transmit RF signal with the desired frequency, the tag receives the signal and supplies to the chip which is attached to it in the transponder. The chip after getting sufficient voltage is able to wake up and hence retransmit the signal at the same frequency to the reader. Thus the purpose of matching an antenna with its load is to insure that maximum power transferred from antenna to chip. To do this, it is needed to have a perfect match between the antenna and the chip. Perfect antenna matching can be achieved by changing the dimension of an antenna, by adding a reactive component or implementing both of them. A mathematical expression can be overviewed as depicted in Figure 3.

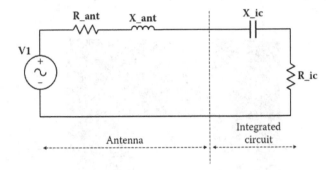

Figure 3. Series model for transponder chip and antenna [3]

The power delivered form antenna to the load or chip is given as [3]:

$$P_l = \frac{R_{ic}}{2\left[(R_{ant} + R_{ic})^2 + (X_{ant} + X_{ic})^2\right]} V^2_{ant} \tag{1}$$

In the above equation it can be seen that the maximum power can be delivered from the antenna to the IC only if $R_{ic} = R_{ant}$ and $X_{ic} = -X_{ant}$. Thus it can be observed that the maximum power can be delivered from antenna to load only if they are conjugate matched. This gives one of the favorable conditions for antenna designer as generally the antenna impedance is inductive in nature and the impedance of the chip is capacitive.

In this research the antenna is designed to work at 869 MHz. At this frequency the input series impedance of the chip is 128-j577 Ω. Thus the requirement is to have antenna impedance of 128+j577 Ω such that it is complex conjugate matched with the load and maximum power is transferred.

Conjugate Match Factor (CMF) is the factor which tells how good matching is done between the chip impedance and the antenna impedance. It can be described as the ratio between antenna input power with given chip impedance Z_s and antenna impedance Z_a assuming Z_a is complex conjugate of Z_s.

The value of CMF changes between 0 and 1 in linear. To receive maximum power from the reader and retransmit the maximum power to the reader, the antenna impedance should be complex conjugate match and equaled with that of chip.

3. H-slot microstrip patch antenna for UHF RFID

In this section the passive UHF RFID tag design is discussed. This RFID tag is textile made and involving the human body as the object to be tagged.

The designed antenna layout is an H-shape slot place onto a patch, Figure 4.

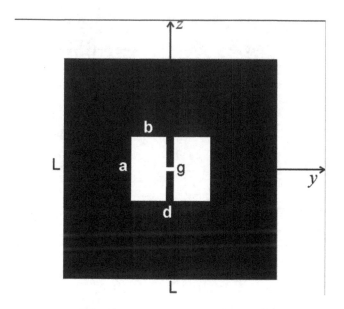

Figure 4. Geometry of nested - slot. The microchip is placed in the central gap of the slot [4]

The patch with H-slot is placed on a substrate and grounded by a conductive material to decouple from the human body. H-slot is a tuning slot for the required conjugate impedance matching between the microchip and the antenna tag. The maximum size of the antenna is 150 mm x 180 mm and the gain is rather poor around -7 dB due to the bidirectional radiation of the slot. But the maximum gain can be increased by increasing width of the tag antenna. And also the impedance matching is done by tuning the internal slot size.

Dielectric material of this patch antenna has a thickness of h and it has a longer face of it in the lower part which is placed on the human body through the conductive ground plane. It is an advantage to have longer ground plane because it will avoid the effect of human body radiation.

The radiation is produced by the patch open edge and by the slot. To achieve better radiation performance, width of the antenna can be increased depending of available place for tag. The dimension of the central gap is kept fix by the microchip packing but for tuning the other dimensions of the slot are optimized. The perfect conjugate matching should be done between antenna and microchip to obtain the maximum reading distance.

4. Observations

4.1. Impact on antenna performance with radiating element having different surface resistivity.

While working with textile antenna, it is found that the antenna is not working properly as it should work. This is because the antenna which is generally made of very high conductive material has very good radiation efficiency and gain. However the antenna made of conductive textile material has very high surface resistivity and hence lower conductivity. Because of this property of textile material it is difficult to choose the appropriate conductive textile material for desired gain. Also the height of dielectric constant plays a major role in determining the radiation efficiency.

Because of the problem of higher surface resistivity associated with the conductive textile material, a relation between surface resistivity, gain and radiation efficiency is analyzed. For this purpose the microstrip patch antenna is simulated in simulation software IE3D. First the measurement is performed by simulating two different microstrip patch antenna. Both of these antennas are simulated to work at a frequency of 2.45 GHz but with the radiating element having different surface resistivity.

The surface resistivity for two different radiating elements was chosen to be 0.02 Ω/sq and 1.19 Ω/sq. Fleece fabric is used as dielectric material which has a dielectric constant of 1.25. The antennas are simulated for reflection loss less than -40 dB and the result is noted.

It is observed from the simulation result that, though all the other antenna parameter are same, the difference in surface resistivity of the two radiating element affect a lot in their radiation efficiency and gain.

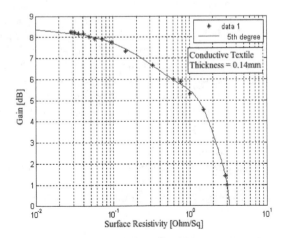

Figure 5. Gain Vs Surface Resistivity Plot

To know the relation between these parameters, a measurement is done for 20 different microstrip antennas keeping other parameters same and only changing the surface resistivity. The result is then plotted in matlab. These measurement results obtained when analyzing different antennas provide valuable information when a conductive textile material is to be used to design an antenna.

Figure 5 depicts that for very low surface resistivity, the gain is maximum. When the radiating element (antenna) surface resistivity is increased, the gain of the antenna starts to decrease.

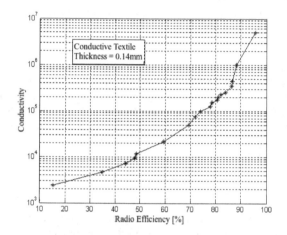

Figure 6. Radiation Efficiency vs. Conductivity Plot

Figure 6 illustrates the relation between the conductivity and the radiation efficiency for above mentioned conductive textile material when used as an antenna. Radiation efficiency is a ratio of power radiated by antenna and power input to antenna. If most of the power input to the antenna is radiated, then the antenna is said to have high radiation efficiency. It can be seen from Figure 7 that conductivity is related to radiation efficiency in logarithmic manner. When the conductivity of radiating element is lower, the radiation efficiency is also very small, and increases as the conductivity increases. However radiation efficiency does not increase in linear way, and to achieve the radiation efficiency in higher percentage, the conductivity of the radiating material should be very high.

The entire simulated antenna has reflection (S_{11}) less than -35 dB. However the entire antenna does not have same S_{11}. This affects the smoothness of the curve obtained in Figure 5 and 6.

4.2. The impact on antenna performance when the thickness of dielectric material is changed to different values.

When the dielectric material thickness is changed, this affects the radiation efficiency. To analyze this effect, three microstrip patch antenna is designed.

Frequency	2.45	GHz
Surface resistivity of radiating element	0.02	Ω/sq
Thickness of radiating element	0.14	mm
Dielectric material	Fleece fabric	-
Dielectric constant	1.25	-
Height of Dielectric material	1, 2, 3	mm

Table 2. Technical specification for the antenna

Three rectangular patch antennas are designed for different height of dielectric material. On doing simulation, various parameters like reflection, gain, radiation efficiency and antenna efficiency for different patch antenna were observed. The obtained results are shown in Figure 7 and 8.

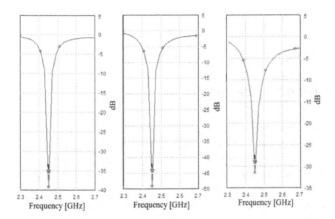

Figure 7. S₁₁ for antenna with dielectric thickness 1 mm, 2 mm and 3 mm correspondingly

From above S_{11} plot it can be seen that the reflection is less than -30 dB for all three antennas with dielectric thickness 1 mm, 2 mm and 3 mm.

For the same specification of antennas, the radiation efficiency is measured with different height of dielectric material.

The above plot gives the measure of radiation efficiency of the antenna with three different thicknesses. It can be seen that for an antenna working at 2.45 GHz and dielectric thickness of 1 mm, the radiation efficiency is 61.2 %, for dielectric thickness of 2 mm, the radiation efficiency is 83.4 % and for dielectric thickness of 3 mm, the radiation efficiency is 90.2 %.

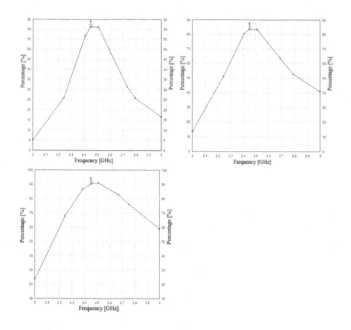

Figure 8. Radiation Efficiency for antennas having thickness 1 mm, 2 mm, 3 mm respectively

5. Conductive textile materials

The fabric that can conduct electricity is called conductive fabric. The conductivity of the fabric depends on how it is manufactured. Conductive fabric can be made in various ways. They can either be produced by metal inter woven fabric during manufacturing or by metal coated fabric [5] also called electro thread. These conductive textiles have wider application in various fields as they are used for shielding human body and some special equipment from external electromagnetic radiation, and also as pressure sensor or flexible heaters, which are made out of easily wearable conductive textile [6].

For a good design of a textile antenna, the conductive fabric should satisfy some of the conditions as given below.

- The electrical resistance of the conductive textile fabric should be small in order to reduce the ohmic losses in the fabric.

- The surface resistivity should be homogeneous over the entire conductive textile fabric i.e. the variation of resistance should be minimum.

- The fabric should be flexible enough to be able to use as a wearable antenna.

The antenna performs better if the conductive textile fulfills the above given characteristics.

Non woven fabric

By the name it can be concluded that non woven fabric is prepared by neither knitting process, nor are woven fabric. Thus the non woven fabric does not go through the initial stage of yarn spinning and also a definite web pattern as that of a woven fabric is not obtained. Non woven fabric manufacturing process is similar to that of paper manufacturing process.

The material used is Cu-Ni with the thickness measured in the lab is 0.14 mm.

Description	copper + nickel plated non-woven polyamide fabric
Roll widths	102 cm ± 2 cm
Surface resistivity	Max average 0,02 Ohm/square
Shielding effectiveness	70-90 dB from 50 MHz to 1 GHz
Purpose	conductive fabric for general use
Temperature range	-30 to 90 (degree centigrade)

Table 3. Technical specification of Cu-Ni textile material [7]

These kind of fabric are generally manufactured in three ways namely, drylaid syste, wetlaid system and polymer based system. After the fabric is manufactured, it is then strengthened. There are various ways for strengthening fiber web as by using chemical means by spraying, coating. This can also be achieved by thermal means by blowing air or by ultrasonic impact which partially fuses (connects) the fiber thread. Thus finally the metal layer is coated.

Woven Fabric

Woven fabric is a construction design for lab use at CTU in Prague. The woven fabric consists of silver nano particles attached to the thread of fiber when being constructed and then woven to form a conductive textile, Figure 9.

Name	Betex
Materials Used	Shiledex (60%), Polyster (40 %)
Number of fiber threads per centimeter	20
Surface Resistivity	1.19Ω /sq.

Table 4. Technical specification of Betex textile material

5.1. Electrical resistance and resistivity of textile materials

Electrical conductivity of textile materials is calculated from electrical resistivity as:

$$\sigma = 1/\varrho \qquad (2)$$

where σ is electrical conductivity [S/m], ϱ is electrical resistivity [Ω m].

Figure 9. Photo of woven Betex textile material

Electrical resistivity can be obtained via resistance mesurement [8-10]. We differentiate surface and bulk electrical resistance. Surface resistance is defined as the ratio of a DC voltage U to the current I_S which flows beteen two specific electrodes. The electrodes are placed on the same side of measured material and it is assumed all currents flows only between electrodes and do not penetrate into the bulk of material [8].

Bulk resistance or electrical resistance takes into account all currents flowing in the material, not only on the surface. It can be measured by RLCG bridge or DC power source (showing voltage and current values).

5.2. Resistance modelling

The textile material can be modelled as finite grid of resistors. However, it assumes only woven fabric [11]. The example is depicted in Figure 10.

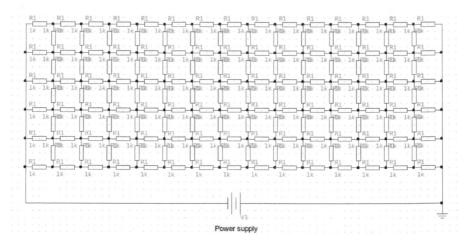

Figure 10. Equivalent circuit diagram of Betex textile material

Measurement of resistance (bulk or surface) is based on placing two square electrodes on two ends of the woven textile material. The structure can be interpreted as series-parallel connection of resistors. The battery represents electrodes and resistors the textile fibres.

The equivalent circuit diagram can be simplified with respect to basic physical laws. Equipotential points in this diagram are in all individual „vertical" resistor connections. The resistors placed between the points with same potential can be eliminated because they are equalled to zero. The voltage probes are placed in the equipotential points in Figure 11. The results are depicted in Figure 12. All probes reach the same value and therefore the resistors can be eliminated.

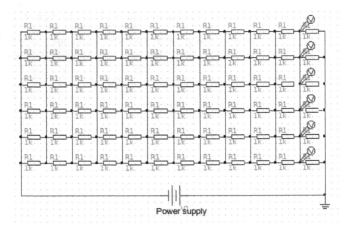

Figure 11. Simplified equivalent circuit model

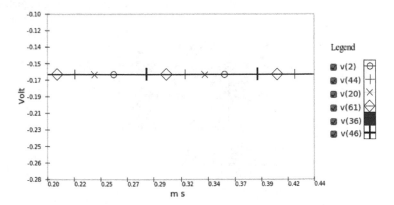

Figure 12. Result values of placed voltage probes.

The resultant resistance of this model can be calculated as series-parallel connection of resistors as:

$$R = \sum_{1}^{12} \frac{R1}{6} = \frac{12\,R1}{6} = 2R1 \tag{3}$$

Formula (3) can be generalized as:

$$R = \sum_{n=1}^{r} \frac{R1}{s}, \ n, r, s \in N \tag{4}$$

where n, r represents number of squares in „horizontal" direction and s in „vertical" direction.

Considering Betex sample and setup measurement, the Betex sample reach dimensions 10 x 3 cm, 25 threads/cm in warp and 20 threads/cm in weft. Parameters n and s are then equalled to:

$$r = 10 \cdot 25 - 1 = 249 \tag{5}$$

$$s = 3 \cdot 20 - 1 = 59 \tag{6}$$

Resultant resistance is equalled to:

$$R = \sum_{n=1}^{r} \frac{R1}{s} = \sum_{n=1}^{249} \frac{R1}{59} = \frac{249R1}{59} = 4.22R1 \tag{7}$$

The parameter $R1$ represents a resistance element of used fiber which forms the whole fabric. It can be calculated from the dimensions of textile structure with the aid of fiber diameter measurement. $R1$ is set to 0.97Ω and $R=4.09\Omega$. It meands the structure is very conductive.

Figure 13. Measurement setup

5.3. Resistance measurement

The Betex sample is measured by RLCG bridge with respect to its calculated resistance. DC power source can cause sample damage at low voltage values (10 V corresponds to approx. 10 A). The measurement setup is depicted in Figure 13.

The measurement of Betex sample shows the resultant resistance is approx. 4Ω which confirms the modelling results.

6. Simulations

In this chapter different two antenna types are designed for the conductive textile material. The two designed antennas are H-slot antennas for RFID having an IC chip EM4222 at 869 MHz with the impedance 128-j577, microstrip patch antenna for RFID having T-match. Manual calculations and simulation results for all the antennas are presented below.

6.1. Simulation of H-slot patch RFID tag antenna

The wearable tags are designed on IE3D, fabricated and tested in real conditions. The overall size of the H-slot antennas is 180 mm x 200 mm. This big dimension of the antennas can be smaller by using a substrate which has a high dielectric constant, because the antenna size depends on the dielectric constant ε_r of the substrate and also the design frequency. In this design a fleece fabric is used as a substrate which has a dielectric constant of 1.25. This fabric is chosen because of its better radiation performance. When a substrate has low dielectric constant and small thickness then the designed antenna has good radiation performance. But if a small tag is requested then a substrate with high dielectric constant can be used.

Three different substrate materials are used for comparison. The first substrate is Polyethylene (ε_r = 2.25, thickness h =1.7 mm). This design gives smaller dimensions (145 mm x 160 mm) then fleece fabric.

Later on this substrate replaces with the fleece fabric to increase the radiation performance because of fleece's low dielectric constant.

Third design is made by a substrate material which has a very high dielectric constant ε_r, to reduce the antenna size. The material is silicone slab. Silicone slab is chosen because it is elastic, hydrophobic material and this property gives an advantage of avoiding the water absorption into the substrate, this is important property of silicone slab because when water absorbed into the substrate, the dielectric constant of the substrate changes. This also gives homogeneous connection between the substrate and radiating patch. The antenna design with silicone slab is giving as a size of 57 mm x 78 mm, much smaller than other designs. But a decreasing in the radiation performance is achieved. Thus there is a trade of between antenna performance and size of the antenna.

In this work, two different conductive textiles are used to design and fabricate two different tags, TAG1 and TAG2.

Parameter	Cu-Ni	BETEX
Frequency	869 MHz	869 MHz
Surface resistivity	0.02 Ω/sq.	1.19 Ω/sq.
Thickness	0.14 mm	0.35 mm
Conductivity	357143 S/m	2381 S/m

Table 5. Specifications for the conductive textile radiating element

Parameter	TAG1	TAG2
Conductive Material	Cu-Ni, conductivy of 357143 S/m	BETEX, conductivity of 2381 S/m
Substrate Material	Fleece Fabric	Fleece Fabric
Thickness of the Substrate	4 mm	4 mm

Table 6. Specification for simulating antenna (TAG1 and TAG2)

6.1.1. Antenna layouts and designs

Cu-Ni conductive textile has high conductivity and the designing with this conductivity gives better radiation performance then Betex textile which has low conductivity. The simulation of two antennas with two different materials having same dielectric substrate and the same dielectric constant is shown in Figure 14.

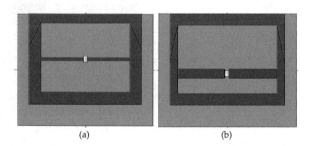

(a)   (b)

Figure 14. Simulated H-slot antenna for radiating element having surface resistivity = 0.02(a), and surface resistivity =1.19(b)

As expected, the tag with high conductive material is giving better radiation performance. When a comparison made between this two tag's simulations, a high reading distance and high radiation efficiency are achieved from TAG1 which is designed with high conductive textile. This can be shown by the plot obtained from the simulation, Figure 15.

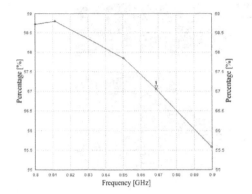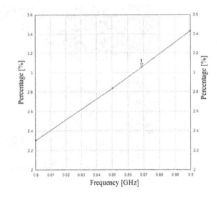

Figure 15. Radiation efficiency Vs Frequency plot for TAG 1 and TAG 2 respectively

It can be seen that radiation efficiency obtained for TAG1 is 57.1% and that for TAG 2 is 3.05%. As it is concluded before that the radiation efficiency is directly proportional to the surface resistivity of the radiating patch. Therefore, higher radiation efficiency is obtained for the TAG having lower surface resistivity. If the antenna were designed from copper material the simulated efficiency would be very high because copper has high conductivity then these conductive textiles. This is shown in later case. Therefore, this value of 57% radiation efficiency is quite good result for this conductive textile.

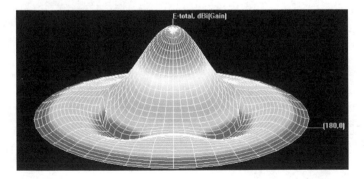

Figure 16. pattern of TAG1

The value of CMF is used to find how good the matching is in between the antenna impedance and chip impedance. The plot obtained for TAG 1 and TAG 2 is depicted in Figure 18.

From the above figure it can be observed that the CMF for TAG1 is 0.975 and that for TAG2 is 0.874. It can be considered that both of the tag antennas has good matching with the chip, however TAG 1 shows better match among the two.

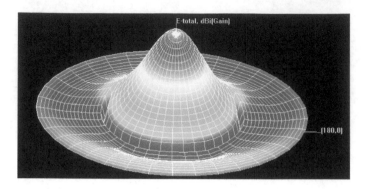

Figure 17. pattern of TAG2

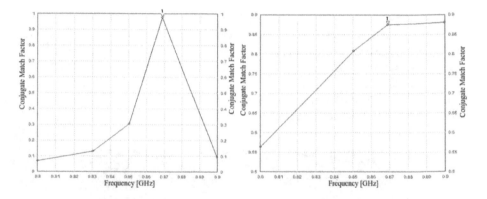

Figure 18. Conjugate match factor plot for TAG 1 and TAG 2 respectively

@869Mhz	TAG1	TAG2
Radiation Efficiency,	57.1%	3.05%
Conjugate Match Efficiency	28.5%	1.52%
CMF	0.975	0.874
Gain	-8.46dBi	-15.25 dBi
Directivity	7.89dBi	8.47 dBi
Size of the Tag Lg x Wg	180x200mm	180x200mm
Antenna impedance, Za	100+j597	206+j472

Table 7. The combined results obtained from the simulation of two tag antenna

6.1.2. Reading range calculation

The read range is obtained as:

$$d_{max} = \frac{c}{4\pi f} \sqrt{\frac{EIRP_R}{P_{chip}} \tau G_{tag}} \tag{8}$$

$$\tau = \frac{4R_{chip}R_a}{|Z_{chip} + Z_a|^2} \le 1 \tag{9}$$

Here, chip sensitivity is -10 dBm and the maximum radiated power by the reader is 3.2 W EIRP. Thus from the formula the transmission power coefficient for TAG1 and TAG2 is equal to,

$$\tau_1 = 0.97 \tag{10}$$

$$\tau_2 = 0.87 \tag{11}$$

Thus the maximum range obtained is $d_{max} = 2$ m for TAG1 and $d_{max} = 1.2$ m for TAG2.

6.2. Comparison when different dielectric substrate used

A comparison is performed in simulation to compare the radiation efficiency of H-slot antenna, when the conductive textile (Cu-Ni) is used as radiating element. The comparison is made by changing the thickness of dielectric substrate (fleece fabric) to 2 mm, 2.56 mm and 4 mm. The simulation is performed for the conductor having the surface resistivity 0.02 Ω/sq to resonate at the frequency 869 MHz. For three different thickness values, three different antenna geometry and CMF and radiation pattern is achieved. The CMF for all the designs were measured to be more than 0.95 when measured in linear scale.

Figure 19 depicts the radiation efficiency obtained from 4 mm thick dielectric substrate is the highest and obtained to be 57.1 %, for 2.56 mm thick dielectric substrate is 43.4 % and that for 2 mm thick dielectric substrate is 35.8 %. The better performance is achieved with the antenna having thicker dielectric substrate.

6.3. Comparison when different dielectric material used

In this case the two dielectric materials are used. One is the fleece fabric with the dielectric constant 1.25 and the other is silicone slab with dielectric constant 11.9. The silicone slab with high dielectric constant is used to reduce the size of antenna and also hydrophobic in nature. This is very useful characteristics of silicon slab. The two antennas with two different materials are depicted in Figure 20 and 21.

As seen from the graph, the radiation efficiency of antenna using silicone slab and having higher dielectric constant, is reduced compared with the one using fleece fabric. Though the

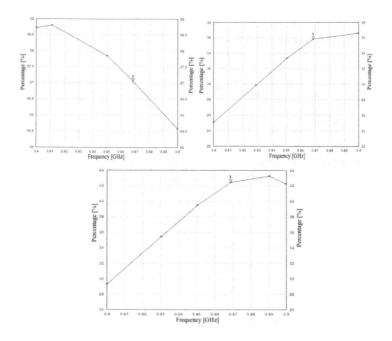

Figure 19. Simulation result of Radiation efficiency plot for Cu-Ni, with the height of dielectric substrate 2 mm, 2.56 mm and 4 mm.

Figure 20. Antenna with Dielectric substrate fleece and antenna with dielectric substrate silicone slab respectively

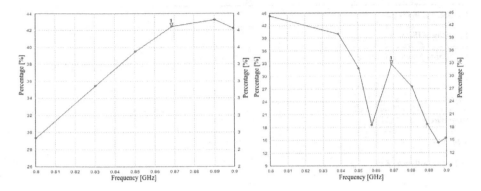

Figure 21. Radiation efficiency of antenna using dielectric substrate fleece (43.4%) and silicone slab (32.7%)

size of antenna was reduced, the efficiency was also decreased drastically. It is due to this reason fleece fabric is preferred.

6.4. Simulation of microstrip patch antenna for RFID application

This is another technique of designing RFID tag antenna. A rectangular parch antenna is used as a tag antenna for RFID. The microstrip patch antenna for RFID is designed for 869 MHz. The manual calculation of microstrip patch is calculated in the similar ways as for the rectangular microstrip patch in chapter 3, however the feeding is different. A T-match is used to match the impedance of the antenna to the chip. The calculated length is 150 mm and the width is 190 mm.

The dielectric material used is Fleece fabric with dielectric constant 1.25 and dielectric height of the substrate 2 mm. The radiating element is simulation of conductive textile material having surface resistivity 0.02 Ω/sq.

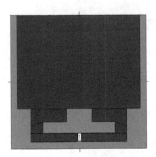

Figure 22. Microstrip patch antenna using T match

Frequency(869 MHz)	TAG3
Radiation Efficiency	41.82%
Conjugate Match Efficiency	20.91%
CMF	0.956
Gain	-7.99dBi
Directivity	8.162dBi
Size LxW	150x190
Imput Impedance , Za	92+j547
Reading distance	2m

Table 8. Simulated output results for TAG 3

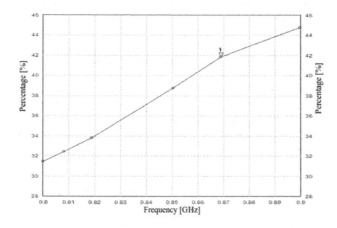

Figure 23. Radiation Efficiency Vs frequency plot for the tag antenna

Figure 23 shows the radiation efficiency obtained is 41.8 %.

7. Fabrication and measurement of H-Slot and patch RFID antenna

7.1. Results of fabrication and measurements of TAG1 and TAG2

For fabrication of these two tag antenna, the available fleece fabric had a thickness of 2 mm, so to have 4 mm thickness two layer of fleece is overlapped by using glue. The conductive material also attached to the substrate by using glue.

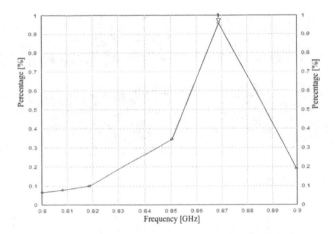

Figure 24. Conjugate match vs frequency for TAG 3 (CMF$_{max}$=0.956)

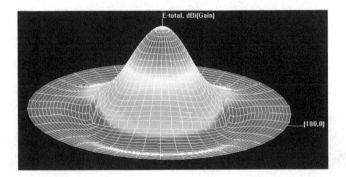

Figure 25. radiation pattern for TAG 3

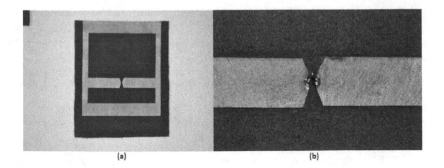

Figure 26. Fabricated antenna (a), chip connection to slot arm (b) for TAG 1

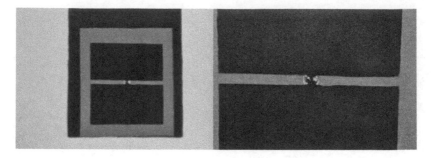

Figure 27. Fabricated antenna for TAG 2 and chip connection

Figure 28. Fabrication and Chip connection for TAG 3

As can be seen from the Figure 27 TAG 2 is constructed from Betex and being very difficult to connect chip by soldering, an alternative way is used. First a copper tape is attached, similar to that with microstrip patch and then the chip is soldered on top of it as shown in Figure. This not done with TAG 1, as it was not necessary.

To measure the tag performance, an RFID reader is connected to the computer.

The reader shown in the above figure generates the frequency signal which is captured by the tag antenna, and retransmit signal back to reader. This signal is received by the reader and is sent to the computer for further processing of the signal.

The RFI21 RFID Reader Demo application program is used in the computer to read the reader. This application uses Python 2.6 programming language in the computer to accomplish this task. The reader is manufactured by METRA BLANSKO a.s.

During the measurement, the reading distance of the TAG1 and TAG2 is measured. Tags are moved towards the reader's antenna till the reader detects the signal from the tag.

Figure 29. Measuring Reading distance of TAG1

As soon as RFID tag is detected by reader's antenna, the information is displayed on the computer.

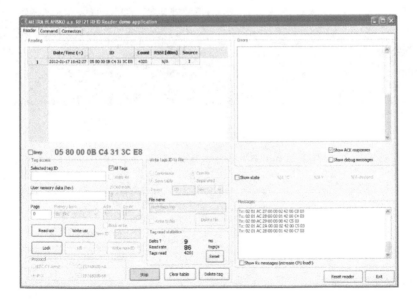

Figure 30. Application program detecting the EM4222 chip ID (in red)

EM4222 chip is used in the tag antenna. When the tag is detected by the reader the tag ID is displayed in the computer. This can be seen in the figure which is the ID of the chip in red color.

From the measurement results, the reading distance for TAG1 is measured to be 50 cm. This is quite smaller then simulated results because of the change in dielectric material due to non-precise determination of permittivity and also soldering process.

The reading distance for TAG2 is quite close the simulation results, 90 cm. The simulated reading distance for this tag is 1.2 m. Thus TAG 2 gave better performance and was very close to simulated values.

Range	TAG1	TAG2	TAG3
Reading range	50 cm	90 cm	60 cm

Table 9. Measured read range from the three designed tag

A short experiment was done to compare working of the fabricated tag and the tag available in the market. To make a comparison of reading distance, two different UHF RFID tag's reading distance were measured. The tags used were UPM Hammer 258-1 and UMP short dipole 211_2. These are the commercially available tag in the market.

(a)   (b)

Figure 31. UPM Hammer 258-1RFID tag (a), UMP short dipole 211_2 RFID tag (b)

Parameter	UPM Hammer 258-1 RFID tag	UMP short dipole 211_2 RFID tag	H-slot TAG2
Reading Distance	98 cm	152cm	90cm
Chip Protocol	EPS S1 Gento	EPS S1 Gento	IP-X

Table 10. Comparison of read range of manufactured tag with commercially available tag

The measurement was performed in open space in the lab.

8. Conclusion

Implementation of textile antennas for RFID tags represents a realistic developmental assignment, and as shown and practically proved, this arrangement yields good results. The successful operation of such textile antennas mainly requires the mechanical stability of the textile composite, which realizes a RFID tag antenna. A good function of the antenna and thus the sensitivity of the complete tag can be provided for just compliance with the mechanical construction and stability of required dimensions. A good choice of textile material for both electrically conductive structures and the insulating layer of the resultant fabric composite is the most important prerequisite for the successful implementation.

Textile RFID tags find its use at both person marking (marking of athletes, protective clothing and other functional ready-made textile products) and stock-in-trade marking in hospitals, packaging, etc.

Acknowledgements

This work was supported by the project Kompozitex FR– TI4/202 - Composite textile materials for humans and technology protection from the effects of electromagnetic and electrostatic fields. The work was conducted in the Department of Telecommunication Engineering at the Czech Technical University in Prague in the scope of thesis called "Design and performance analysis of purely textile antenna for wireless applications" [12] and other research projects.

Author details

Lukas Vojtech, Robi Dahal, Demet Mercan and Marek Neruda

*Address all correspondence to: lukas.vojtech@fel.cvut.cz

Department of Telecommunication Engineering, FEE, CTU in Prague, Prague, Czech Republic

References

[1] Kennedy T.F. et al. Body-Worn E-Textile Antennas: The Good, the Low-Mass, and the Conformal, IEEE Transaction on Antennas and Propagation 2009; 57(4) 910-918.

[2] EM Microelectronic Data Manual, Marina SA. Company of the Swatch Group, Data Sheet EM4222, 2003.

[3] Lozano-Nieto A. RFID Design Fundamental and Application. Boca Raton: CRC Press; 2011.

[4] Occhiuzzi C. et al. Modeling, Design and Experimentation of Wearable RFID Sensor Tag, IEEE Transactions on Antennas and Propagation 2010; 58(8) 2490-2498.

[5] Aniolczyk H. et al., Application of Electrically Conductive Textile as Electromagnetic Shields in Physiotherapy, Fibers and Textiles in Eastern Europe 2004; 12(4) 47-50.

[6] Vojtěch, L. and Neruda, M. Application of Shielding Textiles for Increasing Safety Airborne Systems - Limitation of GSM Interference. In The Ninth International Conference on Networks (ICN 2010). Los Alamitos: IEEE Computer Society 2010, 157-161.

[7] Constantine A.B. Antenna Theory Analysis and Design. New York: John Wiley; 1997.

[8] Maryniak W.A., Uehara T., Noras M. A. Surface Resistivity and Surface Resistance Measurements Using a Concentric Ring Probe Technique, Trek Application Note 2003; 1005: 1–4. www.trekinc.com/pdf/1005_Resistivity_Resistance.pdf (accessed 16 August 2012).

[9] IEC 61340-5-1 Standard. Electrostatics – part 5-1: Protection of Electronic Devices from Electrostatic Phenomena - General Requirements, 2001.

[10] ASTM Standard D 257-99. Standard Test Methods for D-C Resistance or Conductance of Insulating Materials, 1999.

[11] Neruda, M. and Vojtěch, L. Verification of Surface Conductance Model of Textile Materials. Journal of Applied Research and Technology 2012; 10(4), 579-585.

[12] Dahal, R. and D. Mercan. Design and performance analysis of purely textile antenna for wireless applications. Sweden, 2012. 64 p. Diploma thesis. Czech Technical University, University of Gävle, Sweden. hig.diva-portal.org/smash/get/diva2:504525/FULLTEXT01 (accessed 16 August 2012).

Chapter 11

Object and Human Localization with ZigBee-Based Sensor Devices in a Living Environment

Hiroshi Noguchi, Hiroaki Fukada, Taketoshi Mori,
Hiromi Sanada and Tomomasa Sato

Additional information is available at the end of the chapter

1. Introduction

Radio frequency identification (RFID) systems are currently widespread in business applications such as inventory management and supply chain management. In particular,
the active type of system is often used for total management over a large area because of
its long communication range. With the increasing miniaturization and price reduction of
RFID tags, the applicable area is expanding from business to consumer. One of the most
promising areas is the home environment. In the home environment, object management
is as important as that in business areas such as warehouses, because there are many
pieces of equipment in daily use. While many objects are accessed often in a warehouse
environment, only a few objects are replaced in the home environment. The management
of massive objects is thus unnecessary for the home. Another application is in the understanding of the environmental situation. In warehouses, tags with temperature and humidity sensors are often used for quality management. These tags realize not only object
identification but also understanding of the object's situation. The next step in situation
understanding is behavior recognition for home occupants. Behavior recognition is useful
for intelligent home automation, healthcare based on life patterns, and monitoring of
people living in remote locations.

To capture human behavior using an active RFID system, the system must measure various information related to human behavior. A typical example of human behavior measurement using a wireless sensing system, which is regarded as a kind of active RFID
system, is MITes [1]. MITes can capture home environmental information (e.g. lighting
changes and passing people) using wireless sensor devices attached to each of the
rooms. MITes can also measure details of human behavior using wearable sensors. Tradi-

tional active RFID systems can capture large segments of human behavior, even without the use of a complicated system like MITes. Environmental information can be measured using sensors in the tags. With the addition of the environmental information, the RFID system can easily identify the object with the attached tags. However, while identification alone is suitable for object management, information about object handling and object locations is required to measure human behavior. For object handling, the work of Philipose et al. [2] indicates that the object handling sequence assists with the estimation of human behavior. This information can be captured easily with sensors included in the tags. For the location, as an example of the use of location information for human behavior recognition, the information that a cup exists on a sink indicates that someone is washing the cup. The presence of the cup on a table suggests that someone is drinking from it. Also, if it is known that one specific person uses the cup, this information also identifies the person who is drinking. Although direct information about humans is desirable for behavior recognition, direct measurement is difficult with active RFID systems. If the inhabitants wear tags, some information can be captured. However, wearing the tags constricts the natural behavior. Intille et al. [3] suggested that a rough human location is useful for human behavior recognition. Based on their work, we decided that our measurement target for humans using active RFID systems is sub-room-level human localization without the humans wearing tags. Therefore, our research goal is object and human localization using an active RFID system.

Popular approaches for tag position estimation use radio signal strength indicators (RSSIs) for communication between tags and readers, because RSSI depends on the distance between the tag and the reader [4-6]. The simplest approach uses a triangulation algorithm. However, in the home environment, which contains many obstacles for RFID systems such as furniture and electrical appliances, localization is more difficult because the strength of the radio wave can change easily with the room situation. One solution is the deployment of multiple reference tags, which indicate true position [5] [6]. However, this approach is impractical in a living environment because of the cost and difficulty. Distortion of the radio waves by the occupant's presence decreases the localization performance. When we consider the above applications, accurate position (i.e. x-y-z position) estimation is not necessary, but rough location (e.g., on a table, in a drawer, or in a cabinet) is required. Based on this idea, we have already proposed a method for localization of tag-attached objects [7]. The method uses a machine learning technique and a rule-based algorithm to combine RSSI data and sensor data captured by externally distributed sensors across the room. This combination improves the performance in the presence of humans. However, this method has some disadvantages, including the cost of a commercial RFID system, the necessity for the tag readers to have a local area network (LAN) connection, the additional introduction of distributed sensors and the limitations of the estimation locations (e.g. the system cannot distinguish any drawers that do not contain switch sensors).

To overcome these problems, we must use a new active RFID system instead of the current commercial active RFID systems. We have focused on ZigBee technology for wireless communications. ZigBee has advantages for accurate localization. RSSIs in ZigBee are sensitive to distance because of its high frequency radio wave. Another advantage is that ZigBee provides protocols for sensor devices, which leads to easy transmission of the sensor data from the tags. However, because the ZigBee-based RSSI is more sensitive than a low-frequency RFID system, the presence of humans disturbs the RSSI more severely. The use of sensor data on tags would improve the object localization performance. Rowe et al. [8] have already reported that limitation of the location candidates improves the localization performance. We have expanded the previous algorithm to prevent performance degradation. The algorithm uses the RSSI data, the environmental sensor data, and data from the sensors on the tag to prevent degradation of the performance by human interference.

On the other hand, the sensitivity of ZigBee-based RSSIs to the presence of humans is effective for human localization. Wilson and Patwari developed a human tracking method based on RSSI values from reference nodes at the outside of the walls [9]. Their approach requires many wireless devices to generate tomography data for tracking, and no obstacles exist in the room. For our application, we do not need high-resolution human positioning but require only rough location using a few devices. If human interference with the radio waves is stable, the pattern of the RSSI values among the nodes specifies the human location. Our challenge is therefore to estimate a sub-room-level human location based on this RSSI distortion using a fingerprinting approach, which is the same as object localization.

In this paper, we constructed a prototype active RFID system using ZigBee devices. We also proposed an object localization method using RSSIs among tags and data from sensors attached to the tags. Our experimental results demonstrated the feasibility of our localization approach for both objects and persons in a realistic home environment. The results also show that our approach reduces the performance degradation caused by the presence of humans.

2. ZigBee-based sensor device

To avoid limitations in the sensor variety and the communication protocols, we developed a new ZigBee-based prototype system. The system consists of the target nodes, which are tags in the RFID system, and the reference nodes, which are readers in the RFID system. The difference between this system and the traditional RFID system is that our system enables communication among the readers and can gather RSSI data because the reference nodes are also regarded as a kind of target node. The devices consist of the XBee, which is a commercial ZigBee communication module, and the Arduino or Arduino Fio microcontroller, which is commonly used in prototype device construction because of its compactness and ease of programming. The antenna used for wireless transmission and reception is non-directional to reduce the system performance dependence on device direction. The developed sensors and deployment examples are shown in Fig. 1.

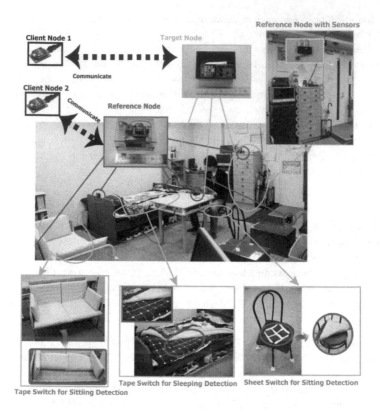

Figure 1. Developed devices and deployment examples.

2.1. Target node

The target node is used for object identification and localization. The node is attached to an object in a room. The node consists of the Arduino Fio and the XBee. The node contains an acceleration sensor (ADXL355) for detection of object handling, along with a luminosity sensor (CdS cell), a humidity sensor (HIH-4030), and a temperature sensor (TEMP6000) for environmental status measurement near the object. The node is battery powered. However, the current device has a battery life of only 3 days, and provision of longer battery life will be part of our future work.

The target node detects the object handling state by using an acceleration sensor, which acts as a trigger to localize the object position. In our research, we estimate the following five motion states by analysis of the acceleration changes:

i.   Stable: object is in a stable state;

ii.   Start Moving: object begins to move;

iii.   Keep Moving: object continues to move;

iv.   Ambiguous: object is either in "Moving" state or in "Stable" state;

v.   Stop Moving: object stops moving.

To be specific, when a node shows noticeable changes in acceleration beyond a set threshold after a long time in the "Stable" state, our system judges this change to be to the "Start Moving" state. Then, as long as the acceleration sensor continues to respond, the state is regarded as being the "Keep Moving" state. However, in the real case, even if an object is moving, the acceleration sensor attached to the node sometimes does not show any noticeable response because of the way it moves. To avoid mistaken estimation in such cases, where even changes in acceleration cannot be detected, the system does not instantly determine the state to be "Stop Moving". Instead, the system regards such a state as "Ambiguous", which means that the node is either in the "Keep Moving" state or the "Stop Moving" state. If the acceleration sensor does not output any noticeable changes after a fixed period of time, the system decides that the first moment where the acceleration sensor's response disappears is the "Stop Moving" state, and the subsequent moments are the "Stable" state. Typical detection results using this algorithm are shown in Fig. 2.

To examine the validity of this algorithm, we performed some preliminary experiments. Because it is difficult to generalize all possible patterns of object motion, in the preliminary experiments, we simply raise an object with a node and move it for a time, and then set it down somewhere. However, despite the simplicity of the algorithm, the system can distinguish the state of object motion from the other states quite well, with a success rate of more than 90% according to our experimental results.

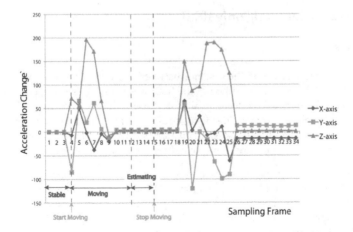

Figure 2. Motion sensing example results with acceleration sensor.

2.2. Reference nodes

A reference node is used for communication with the target nodes and for collection of the environmental sensor data. The node consists of an Arduino and an XBee. The node is capable of connecting to various sensors for environmental data collection. In our experiment, the node contains the same sensors as the target nodes and switch sensors to detect human behavior such as sitting and sleeping. Because the reference nodes cannot move if they are to provide localization reference data, the nodes are attached to fixed objects such as furniture and electrical appliances. The electric power is supplied to these nodes by a power line, because they do not move.

2.3. Communication protocol

The computer for object and human localization collects and controls all the sensor data and the RSSI values. For synchronization and simultaneous data collection, the computer controls the targets and the reference nodes separately with two gateway nodes, which are called client nodes. A typical communication example is shown in Fig. 3. In the figure case, the target nodes and reference nodes transmit sensor data periodically. The reference nodes also regularly gather RSSI values between the reference nodes to estimate human presence and human location based on the algorithm given in section 5 of this paper. When the target node detects object handling using the acceleration sensor, the node transmits a signal to indicate the handling of the object by the occupant. After transmission, the node sends the state of the target node periodically. When the node detects that the object has been put down somewhere, the node broadcasts the putting down action to all reference nodes. Finally, the target node receives each reference node's data with RSSI values and transmits all data to the client node. The computer calculates the object location from the collected RSSIs.

3. Object localization using only RSSI

3.1. Object localization method

While RSSI has a dependence on the distance between the nodes, the RSSI values do not change linearly with the distance. Although the RSSI is sensitive to some types of environmental noise, an RSSI from a fixed location almost always indicates the same value, regardless of the time. Therefore, our main idea is to reduce the environmental effects on the RSSI by not using just a single RSSI, but by using a pattern extracted from several RSSIs. To realize this idea, we must introduce three kinds of pattern recognition method.

The three kinds of pattern recognition method used in our work are the k-nearest neighbor (KNN), the distance-weighted k-nearest neighbor (DKNN) [10], and the three-layered neural network (NN) algorithms. KNN is a method for classification of objects based on the closest training examples in the feature space. The nearest neighbor algorithm, which means that K equals 1, has strong consistency results. As the amount of data approaches infinity, the algorithm is guaranteed to yield an error rate that is no worse than twice the Bayes error rate, which is the minimum achievable error rate given the distribution of the data. KNN is guaranteed to approach the Bayes error rate, for some value of K. DKNN is an extension of KNN, which weights the contributions of the neighbors, so that the nearer neighbors contribute more to the average than the more distant neighbors. We use the inverse of the squared Euclidean distance as a weight function. NN is a kind of classification technique. It is known that NN can demonstrate high discrimination ability for data that has multiple dimensions and is linearly inseparable. We therefore adopted these three methods in our work for object location estimation with RSSIs.

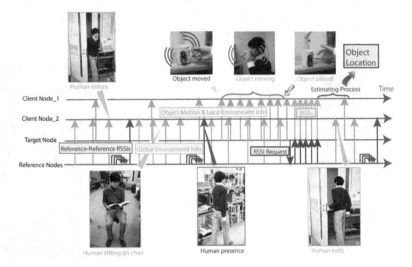

Figure 3. Communication protocol overview.

3.2. Experimental conditions

To investigate the basic object localization performance of the ZigBee-based RFID system, we conducted three experiments. Generally speaking, the classification performance depends heavily on the parameters used in the pattern recognition algorithm. For example, the performance of KNN or DKNN is dependent on the parameters such as the value of k, whereas the performance of the NN depends on parameters such as the number of nodes in the hidden layer. In our experiments, we tried various cases by varying the parameter values and chose the best combination of the parameters according to the estimation performance.

The experimental environment and conditions are shown in Fig. 4. The room contains various articles of furniture. Generally speaking, the largest contributors to reduced localization accuracy are environmental obstacles such as furniture made of metal. This environment provides extreme conditions for localization. However, the difficulty in localization using RSSIs in this environment helps to show that our proposed method is valid in actual living spaces.

To evaluate our estimation algorithm based on pattern recognition methods, we conducted experiments under different conditions: 1) estimation with different numbers of learning data; 2) estimation of different numbers and types of locations; and 3) estimation using different numbers of reference nodes. We collected the same number of RSSI data sets (about 50 to 150) from each of the 17 labeled locations as data sets. The parameters for each of the pattern recognition methods were tuned in advance with the data sets. For performance evaluation, we calculated the estimation accuracy, which is the rate of true positives among the total number of data sets. Ten-fold cross-validation was performed to eliminate any data bias.

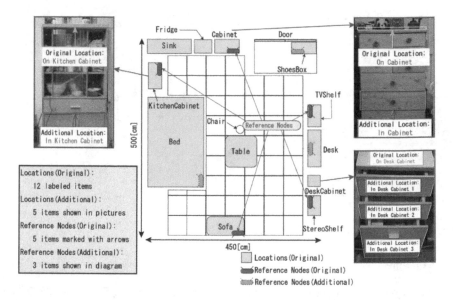

Figure 4. Experimental conditions for object localization using only RSSI.

3.3. Experimental results

3.3.1. Estimation with different numbers of learning data

As mentioned above, we collected RSSI data at each location in the environment and used these data sets to classify objects into particular locations. In Fig. 5a), "n" indicates the number of RSSI data sets collected at each location.

The graph of the results suggests two things to us. The first is that 50 learning data sets per location are sufficient for localization. Therefore, in the following experiments, we used a learning database that contains 50 data sets per location. The other is that as long as the system uses KNN or DKNN as the pattern recognition method, the estimation accuracy is not so heavily dependent on the number of learning data. However, 3-layered NN has increasing difficulty in estimating the object location as the number of learning data increases.

The estimation accuracy at each location with the 3-layered NN is shown in Fig. 5b). The graph demonstrated that the "Table" seems difficult to estimate with the NN. This is because the table is located right in the middle of all of the reference nodes, which means that the table is far from every reference node.

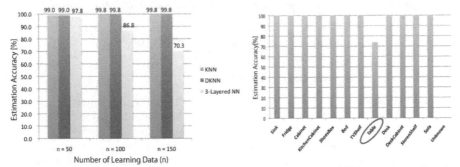

a) Relationship between the number of learning   b) performance of NN at n = 50 per location
    data and estimation accuracy

Figure 5. Experimental results with regard to the number of learning data.

3.3.2. Estimation with different numbers and types of locations

We conducted another experiment to investigate how well our proposed method can accommodate an increase in the number of locations. In this experiment, we added 5 new locations, shown in Fig. 4, to the existing 12 locations. We used a learning database consisting of 50 training data sets for each location and 5 reference nodes to measure the RSSIs with the target node.

Figure 6a) shows that our proposed method can estimate object location effectively even when the number of locations increases. In particular, it has been proved that estimation with KNN and DKNN is hardly affected by an increase in the number of locations, whereas estimation with the 3-layered NN becomes worse when the variety of locations increases. In Fig. 6b), we can see a similar tendency to that which appears in Fig. 5b). However, in this case, the estimation accuracy of the "InCabinet" state also drops seriously along with that of the "Table" state. The reason for this phenomenon is thought to be that it is becoming increasingly difficult to distinguish the "OnCabinet" state from the "InCabinet" state.

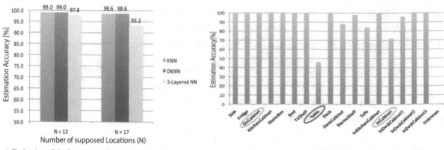

a) Relationship between the number of supported locations and estimation accuracy

b) Performance of NN at n = 17 per location

Figure 6. Experimental results with regard to the number of estimation locations.

3.3.3. Estimation by different numbers of reference nodes

We conducted another experiment to investigate the influence of the number of reference nodes on the estimation accuracy. All the experiments above used the 5 reference nodes illustrated in Fig. 4. In this experiment, however, we changed the total number of reference nodes in two ways: one was to subtract two reference nodes from the existing nodes, and the other was to add three nodes to the existing nodes. In the former case, we only used the reference nodes installed at the kitchen cabinet, the TV shelf, and the sofa, while in the latter case we attached reference nodes to the bed, the shoebox, and the desk.

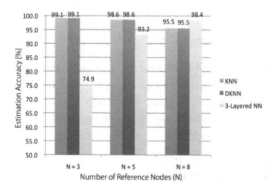

Figure 7. Experimental results with regard to the number of reference nodes.

The graph of the results shown in Fig. 7 suggests two things to us in particular. The first is that there is little difference in the best estimation accuracy with the different numbers of reference nodes. The other is that KNN and DKNN perform strongly with fewer reference nodes, whereas the 3-layered NN has trouble in estimating object locations with fewer reference nodes, although it demonstrates better ability than KNN and DKNN when the number of reference nodes increases.

These results indicate that the ZigBee-based RFID system has the capability for object localization using pattern recognition methods under human-absent conditions.

4. Object localization under human presence conditions

4.1. Object localization using sensor data on target node

The conditions for previous experiments are far from realistic. In a living environment, humans are present and handle the tagged objects. The existence of a human degrades the localization performance because the human body disturbs the radio waves. Our system can measure not only the environmental sensor data but also the sensor data on the target nodes. We extended our previous method [7] to be able to handle the sensor data on the target nodes. Because the previous method limits the location candidates based on estimated human behavior, location candidates are also limited in the new method based on sensor data on target nodes.

In the following algorithm, we use DKNN for RSSI-based localization. Because the sensor data on the target nodes indicates the node location well, the algorithm merges the sensor data on the target nodes into the RSSI-based localization results before combination of the sensor data on the reference nodes.

4.1.1. Integration of target-attached sensor data and rssi-based estimation results

In our system, each target node contains a humidity sensor, a temperature sensor, and a luminous intensity sensor. The humidity and temperature sensors show changes only at specific locations, whereas the luminous intensity sensor is highly sensitive to the environment. This is why the system changes the estimation priority relative to the sensors that have reacted. First, the system integrates the estimation based on humidity or temperature sensors into the RSSI-based estimation. Then, the system integrates the estimation based on the luminous intensity sensor into the results.

• Integration of Humidity and Temperature Sensor Data

Because both the humidity sensor and the temperature sensor change dramatically only at specific locations, the system gives top priority to estimations based on these sensors. For example, because the system can detect object motion through the acceleration sensor, if the humidity rises around the time when an object is set down, it probably indicates that the object has been placed near the sink, because the sink is the only place that can cause a dramatic change in humidity. In the same way, if the temperature drops around the time when an object is set down, it suggests that the object has probably been placed inside the refrigerator, because the preliminary experiments indicate that the temperature only changes dramatically in the refrigerator. The system places its highest level of trust in these sensor reactions because they limit the object location candidates to one in each case. A localization example based on this policy is shown in Fig 8a).

- Integration of Luminous Intensity Sensor Data

The luminous intensity sensor does not limit the object location candidates to only one. This sensor can provide the system with several candidates for the object location. For example, if the luminous intensity drops dramatically around the time when an object is set down, it suggests to the system that the object has been placed in a dark place, such as the inside of a drawer or underneath the bed. Because the luminous intensity changes sensitively depending on the location, the system may even be able to tell the difference between the inside of a drawer and underneath the bed by comparing the sensor's outputs. A localization example based on this policy is illustrated in Fig. 8b).

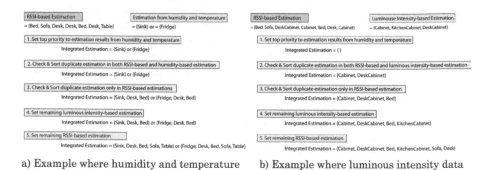

a) Example where humidity and temperature data limits location candidates

b) Example where luminous intensity data limits location candidates

Figure 8. Typical examples of sensor data integration on target nodes.

4.1.2. Integration of sensor data on reference nodes into the results

Because the reference-attached sensor data provide the system with information about human behavior and locations, the system can limit the object location candidates. For example, if a sensor embedded on a sofa continuously reacts around the time when an object is set down, it is easy for the system to guess that the object location is not far from the sofa. In our experimental room, the reference-attached sensors consist of pressure-type switch sensors and microswitch sensors. Pressure-type sensors are installed in the chairs, the sofa, and the bed, whereas the microswitch sensor is installed in the drawer of a cabinet. Each time that an object is set down, the system refers to the reactions of all types of reference-attached sensors around that moment, and keeps track of them. The pressure-type switch sensors, such as those in the chair modules, usually continue to react, not only at the moment when the object is placed, but also during the periods before and after placement, so there is little possibility that the system will fail to detect them. For the microswitch switch sensors such as the drawer modules, however, the sensor reactions usually occur ahead of the moment when the object is placed. If the system only refers to the sensor data within a particular period, it might fail to detect them. However, by tracking the sensor reactions over longer periods, the possibility of missed detection decreases. Thus, the system can use the reference-attached sensors to provide several location candidates, and with the following integration algorithm, shown in Fig. 9, the system integrates reference-based estimation into target-based estimation.

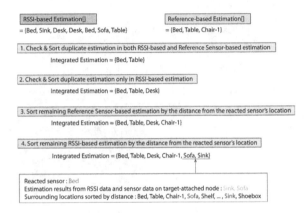

Figure 9. Typical example of sensor data integration on reference nodes.

4.2. Experiment

To evaluate the performance of our system, an experiment was conducted. The experimental room, the sensors for the reference nodes and the deployment locations are the same as those in Fig. 1. The target locations are illustrated in Fig. 10. The total number of target locations is 19. For the training data sets, we collected 400 samples per target location in advance under human absent conditions. For the evaluation, the subject puts down and picks up the object at all of the target locations 5 times, which means 19×5=95 location test data were collected. Strictly speaking, in a single trial, one subject conveyed a target node from location to location in the following order: OnDeskCabinet, InDeskCabinet, StereoShelf, Sofa, Shelf, BedHead, BedBottom, OnKitchenCabinet, InKitchenCabinet, InCabinet, Table, Desk, Chair1, Chair2, TVShelf, ShoeBox, OnCabinet, Fridge, and Sink. For the performance evaluation, we calculated the estimation accuracy in the same way as in the previous experiments. We compared the following five conditions.

1. **RSSI Only:** Estimation based on RSSI data between target node and reference nodes only;

2. **RSSI & Target Sensors:** Estimation directly based on RSSI and target sensor data;

3. **Integration of RSSI and Target Sensor Data:** Estimation based on proposed integration algorithm using the RSSI and target sensor data;

4. **Integration of RSSI and Reference Sensor Data:** Estimation based on proposed integration algorithm using the RSSI and reference sensor data;

5. **Integration of RSSI and All Sensor Data:** Estimation based on proposed integration algorithm using the RSSI and all sensor data (our system performance).

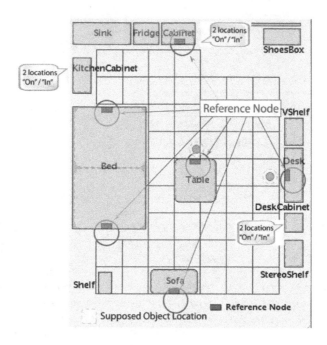

Figure 10. Experimental conditions for object localization with human presence.

The estimation results are shown in Fig. 11. Dynamic interference sources such as a human being had a serious effect on the RSSI-based estimation results. When we use the datasets in our learning database to conduct cross validation, the estimation accuracy is more than 90%. However, in this case, estimation based only on RSSI produced a poor performance.

Estimation based on the RSSI and target sensor data shows lower performance than that of RSSI-only based estimation. In this evaluation, we added another two dimensions (humidity and luminous intensity) to the original RSSI datasets. Because the luminous intensity changes are quite sensitive to the surroundings and to how the target node is placed, they might mislead the estimation to the wrong locations. However, this approach has one point of focus. In the RSSI-based approach, the sink is one of the most difficult places to estimate because it is surrounded by metal. However, by introducing the humidity data, the system estimated the sink correctly through all the scenario tests. This fact indicates that if we integrate target sensor data into RSSI more effectively, then the performance will become higher than that of this simple combination of RSSI and target sensor data.

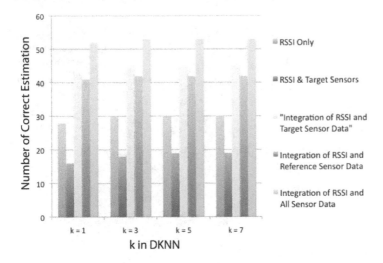

Figure 11. Results for the object localization algorithm at each K in DKNN.

Our proposed integration algorithm based on the RSSI and target sensor data shows much higher performance than the previous two algorithms. It clearly shows the effectiveness of our integration algorithm, which can correct the estimation even if the RSSI-based estimation provides a wrong result. Our proposed integration algorithm based on the RSSI and reference sensor data also shows high performance, similar to that of the RSSI and target sensor data approach. To investigate this in more detail, the contribution of the integration of the reference sensor data is seen to be different from that of the integration of the target sensor data. This therefore indicates that our system should produce a higher performance than these two integration algorithms. The system that integrates RSSI with all kinds of sensor data actually shows the highest performance.

The details of the estimation based on each approach are shown in Fig. 12. These results demonstrate that the use of sensors and limitation of the candidates improves the object localization. The locations where the performance improved are the sinks, the drawers, the bed and the sofa, i.e. locations where the sensor can easily localize the object. These improved locations indicate the effectiveness of the sensor data use. The results also showed that there are several locations that could not be correctly estimated by any of the five algorithms. Any of the five algorithms can estimate an object location based on the results of RSSI-based estimation, but if the RSSIs are heavily distorted by the presence of a human being, even the integration algorithm can hardly correct the mistaken estimation.

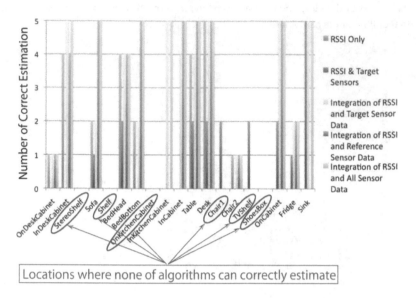

Figure 12. Performance results for each location.

5. Human localization with RSSIs between reference nodes

The human interference with the RSSI values degrades the localization performance. However, because human distortion of the RSSI values is stable, this distortion may be used to indicate the human's location. The human location is estimated by the same approach as that for object localization, using the pattern recognition technique. This is our idea for sub-room-level human localization. While the object location is estimated at the application request time, the human position is always required. Because the continuous use of a target node reduces the battery life, only the reference nodes are used for human localization.

5.1. Experiment on human localization in four areas

To confirm that the distorted RSSI can be used for human localization, we conducted a simple experiment. In this experiment, we make one person stand or sit to cut off the RF signals between two reference nodes. Because this situation drastically disturbs the RSSIs, estimation of the human's location should be easy.

The conditions for the evaluation experiment are described in Fig. 13. The datasets were gathered from 4 reference nodes. When one node is selected to be the base node, as shown in Fig. 13, the node collects RSSI from the three surrounding nodes. In total, 12 (=4×3) RSSIs were used for human localization. In the experiments, the subject sat or stood at the four locations illustrated. Data for the human absence case were also collected. This problem is regarded as 5-class classification. Direct human interference means that the RSSIs between two particular reference nodes are frequently missed, which means that the RF signals could not be received successfully. This data deficit may lead to human location estimation failure. We therefore compensated the part with the data deficit using the average of the successfully collected RSSIs. We evaluated the ratio of the true positive value in all data. Each pattern recognition method adopted the most suitable parameters for the estimation. Ten-fold cross-validation was also used for the evaluation.

The estimation results are shown in Table 1. These results indicate the possibility of estimating the four assumed human locations using the RSSIs among the reference nodes. Estimation with the 3-layered NN algorithm appears to be a little difficult, but estimations based on KNN and DKNN showed high accuracies.

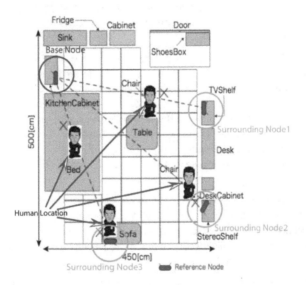

Figure 13. Conditions for experiments on human localization in four areas.

The estimation results suggest two points in particular. The first is that the 3-layered NN algorithm is poor at distinguishing the human presence case from the human absence case. It is also weak at human location estimation when compared with the other two pattern recognition methods. The other point is that KNN and DKNN can not only tell the difference between the human presence case and human absence case, but can also estimate human locations with high accuracy even under the condition where the human presence is unknown.

	Absence	Sofa	Bed	Table	Desk	Total
KNN	96.7%	96.7%	93.3%	91.7%	93.3%	94.3%
		(98.3%)	(88.8%)	(100.0%)	(96.7%)	(95.8%)
DKNN	98.3%	96.7%	93.3%	93.3%	96.7%	95.7%
		(98.3%)	(93.3%)	(100.0%)	(95.0%)	(96.7%)
NN	36.7%	90.0%	95.0%	51.7%	86.7%	72.0%
		(98.3%)	(98.3%)	(68.3%)	(95.0%)	(90.0%)

*Upper selection: Estimation including human absence data.

Lower selection: Estimation excluding human absence data.

Table 1. Results for human localization in four areas with RSSIs among the reference nodes.

5.2. Experiments on sub-room-level human localization

The previous experimental results showed that the direct human interference in communication between two reference nodes contained rich information for human localization. We now address a more complicated case.

The conditions for the evaluation experiment are shown in Fig. 14a). The reference node installed at the table is regarded as the center node, which then receives 14 RSSIs from the remaining surrounding nodes. For human location estimation, we took measurements with each node acting as the center node in turn to cover the whole environment. However, in this case, the same approach will increase the dimensions of the input RSSI data dramatically, which definitely results in the estimation time being too long. Therefore, we only use the reference node on the table as the center node because it is located at the center of the environment and, as Fig. 14a) illustrates, the RSSIs between this center node and other surrounding nodes can cover the majority of the environment.

We divided the environment into 49 grids (0.5 m×0.5 m) as shown in the left part of Fig. 4. We asked a subject to stand or sit on each grid to collect data sets for human localization. Also, human absence was appended to the data sets as one of the conditions. Thus, the problem is regarded as 50 (49+1) class discrimination from 14-dimensional vector data. For the experiment, 50 data sets were collected per location.

In this experiment, we assume an "Unknown" class in the output classes, which is the class to be used when the estimated result is less probable. This means that when the similarity between an input dataset and the most likely dataset in the learning database is smaller than a certain threshold, the system regards the estimated result as wrong and classifies it into the unknown class.

The estimation results are shown in Fig. 14b). The estimation accuracy as a whole is 86.2%, and the discrimination between the human presence case and the human absence case can be discriminated completely, with an accuracy of 100%. The percentage that was estimated as being in the unknown class was 1.3%, which means that almost all of the data is correctly classified.

The results show that RSSIs among the reference nodes can be used as good indicators to localize a human in the environment. It is interesting that although a human standing at the right lower corner of the room does not disturb the radio wave directly, the method estimates the location accurately, which may indicate that the pattern recognition method is sensitive to slight differences caused by human interference.

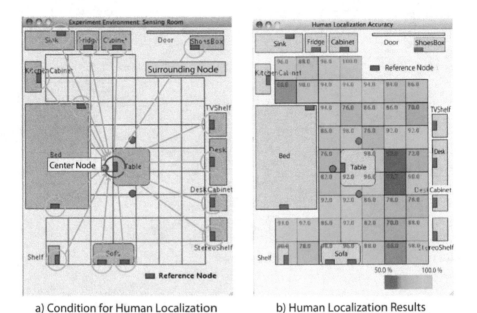

a) Condition for Human Localization   b) Human Localization Results

Figure 14. Conditions and results for the human localization experiment

There are some positions that are difficult to estimate with our approach. The worst two estimations are illustrated in Fig. 15. These scattered estimation candidates are the minority of the estimation as a whole and the majority of the mistaken estimation candidates are quite close to the correct location. This result means that loose conditions such as large grid size may improve the localization performance. Human localization using only RSSIs may contain some trade off between spatial resolution and localization accuracy.

These results demonstrated that the system could localize human positions in indoor environments with RSSIs only.

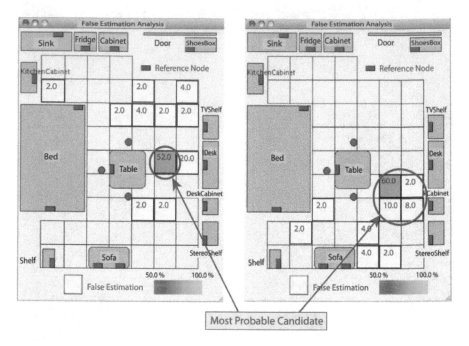

Figure 15. Estimation results at the worst estimation score areas.

6. Conclusion

We proposed methods for object and human localization using a ZigBee-based RFID system. Our method estimates the node locations using a pattern recognition technique from RSSI data among the nodes, environmental sensor data and estimated human behavior to reduce performance deterioration caused by human interference with radio waves. The experiments demonstrated that our method increases object localization accuracy by about 20% under human presence conditions. Considering the fact that human interference with the RSSI is stable, we also performed human localization using pattern recognition based on the RSSI values. Our experimental results showed that our approach is feasible for sub-room-level human localization.

Author details

Hiroshi Noguchi[1*], Hiroaki Fukada[2], Taketoshi Mori[3], Hiromi Sanada[4] and Tomomasa Sato[5]

*Address all correspondence to: hnogu-tky@umin.ac.jp

1 Department of Life Support Technology (Molten), Graduate School of Medicine, The University of Tokyo, Japan

2 Department of Mechano-Informatics, Graduate School of Information Science and Technology, The University of Tokyo, Japan

3 Department of Life Support Technology (Molten), Graduate School of Medicine, The University of Tokyo, Japan

4 Department of Gerontological Nursing/Wound Care Management, Graduate School of Medicine, The University of Tokyo; Tomomasa Sato, Department of Mechano-Informatics, Graduate School of Information Science and Technology, The University of Tokyo, Japan

5 Department of Mechano-Informatics, Graduate School of Information Science and technology, The University of Tokyo, Japan

References

[1] Tapia E, Intille S, Lopez L, Larson K. The design of a portable kit of wireless sensor for naturalistic data collection: proceedings of PARVASIVE 2006, pp. 117--134, 2006.

[2] Philipose M, Fishkin K, Perkowitz M, Patterson D, Fox D, Kautz H, and Hahnel D. Inferring activities from interactions with objects. IEEE Pervasive Computing Magazine 2004;3(4) 10-17.

[3] Intille S, Larson K, Tapia E, Beaudin J, Kaushik P, Nawyn J, Rockinson R. Using a live-in laboratory for ubiquitous computing research: proceedings of Parvasive 2006, pp. 349—365, 2006.

[4] Hightower J, Borriello G, Want R. SpotON: An Indoor 3D Location Sensing Technology Based on RF Signal Strength., Technical Report UW-CSE 2002-02-02, The University of Washington, February 2000.

[5] Ni L, Liu Y, Lau Y, Patil A. LANDMARC: Indoor Location Sensing Using Active RFID: proceedings of the First IEEE International Conference on Pervasive Computing and Communication, pp. 407-415, 2003.

[6] Shih S, Hsieh K, Chen P. An Improvement Approach of Indoor Location Sensing Using Active RFID: proceedings of International Conference on Innovative Computing Information and Control 2006, pp. 453-456, August 2006.

[7] Fukada H., Mori T, Noguchi H, Sato T. Use of active RFID and environment-embedded sensors for indoor object location estimation. Deploying RFID - Challenges, Solutions, and Open Issues, InTech,pp.219-36, 2011.

[8] Rowe A, Starr Z, Rajkumar R. Using micro-climate sensing to enhance rf localization in assisted living environments: proceedings of International Conference on System, Man and Cybernetics, pp. 3668--3675, 2007.

[9] Wilson J, Patwari N. See-through walls: Motion tracking using variance-based radio tomography networks. IEEE Transactions on Mobile Computing, 2011;16(5) 612-621.

[10] Pao T, Cheng Y, Yeh J, Chen Y, Pai C, Tsai Y. Comparison Between Weighted D-KNN and Other Classifiers for Music Emotion Recognition; proceedings of International Conference on Innovative Computing Information and Control 2008, pp. 530, 2008.

Permissions

The contributors of this book come from diverse backgrounds, making this book a truly international effort. This book will bring forth new frontiers with its revolutionizing research information and detailed analysis of the nascent developments around the world.

We would like to thank Mamun Bin Ibne Reaz, for lending his expertise to make the book truly unique. He has played a crucial role in the development of this book. Without his invaluable contribution this book wouldn't have been possible. He has made vital efforts to compile up to date information on the varied aspects of this subject to make this book a valuable addition to the collection of many professionals and students.

This book was conceptualized with the vision of imparting up-to-date information and advanced data in this field. To ensure the same, a matchless editorial board was set up. Every individual on the board went through rigorous rounds of assessment to prove their worth. After which they invested a large part of their time researching and compiling the most relevant data for our readers. Conferences and sessions were held from time to time between the editorial board and the contributing authors to present the data in the most comprehensible form. The editorial team has worked tirelessly to provide valuable and valid information to help people across the globe.

Every chapter published in this book has been scrutinized by our experts. Their significance has been extensively debated. The topics covered herein carry significant findings which will fuel the growth of the discipline. They may even be implemented as practical applications or may be referred to as a beginning point for another development. Chapters in this book were first published by InTech; hereby published with permission under the Creative Commons Attribution License or equivalent.

The editorial board has been involved in producing this book since its inception. They have spent rigorous hours researching and exploring the diverse topics which have resulted in the successful publishing of this book. They have passed on their knowledge of decades through this book. To expedite this challenging task, the publisher supported the team at every step. A small team of assistant editors was also appointed to further simplify the editing procedure and attain best results for the readers.

Our editorial team has been hand-picked from every corner of the world. Their multi-ethnicity adds dynamic inputs to the discussions which result in innovative outcomes. These outcomes are then further discussed with the researchers and contributors who give their valuable feedback and opinion regarding the same. The feedback is then collaborated with the researches and they are edited in a comprehensive manner to aid the understanding of the subject.

Apart from the editorial board, the designing team has also invested a significant amount of their time in understanding the subject and creating the most relevant covers. They scrutinized every image to scout for the most suitable representation of the subject and create an appropriate cover for the book.

The publishing team has been involved in this book since its early stages. They were actively engaged in every process, be it collecting the data, connecting with the contributors or procuring relevant information. The team has been an ardent support to the editorial, designing and production team. Their endless efforts to recruit the best for this project, has resulted in the accomplishment of this book. They are a veteran in the field of academics and their pool of knowledge is as vast as their experience in printing. Their expertise and guidance has proved useful at every step. Their uncompromising quality standards have made this book an exceptional effort. Their encouragement from time to time has been an inspiration for everyone.

The publisher and the editorial board hope that this book will prove to be a valuable piece of knowledge for researchers, students, practitioners and scholars across the globe.

List of Contributors

Ramiro Sámano-Robles and Atílio Gameiro
Instituto de Telecomunicações, Campus Universitário, Aveiro, Portuga

Neeli Prasad
Center for TeleInfrastruktur, Department of Electronic Systems, Aalborg University, Aalborg, Denmark

Michel Simatic
INF Department, Télécom Sud Paris, Évry, France

Ming-Shen Jian
Deptartment of Computer Science and Information Engineering, National Formosa University, Taiwan, R. O. C

Jain-Shing Wu
Deptartment of Computer Science and Engineering, National Sun Yat-sen University, Taiwan, R. O. C.

Pierre-Henri Thevenon and Olivier Savry
Léti, Minatec, CEA Grenoble, France

Andreas Loeffler
Chair of Information Technology (Communication Electronics) Engineering, Friedrich-Alexander-University of Erlangen-Nuremberg, Erlangen, Germany

Heinz Gerhaeuser
Fraunhofer Institute for Integrated Circuits IIS, Erlangen, Germany

Muhammad Mubeen Masud and Benjamin D. Braaten
North Dakota State University, Fargo, U.S.A.

Steffen Elmstrøm Holst Jensen and Rune Hylsberg Jacobsen
Aarhus University School of Engineering, Denmark

Ahmed Toaha Mobashsher and Rabah W. Aldhaheri
Department of Electrical and Computer Engineering, Faculty of Engineering, King Abdulaziz University, Jeddah, Saudi Arabia

T. S. Chou
Computer Science Department, University of California at Irvine, Irvine, CA, USA

J. W. S. Liu

Lukas Vojtech, Robi Dahal, Demet Mercan and Marek Neruda
Department of Telecommunication Engineering, FEE, CTU in Prague, Prague, Czech Republic

Hiroshi Noguchi
Department of Life Support Technology (Molten), Graduate School of Medicine, The University of Tokyo, Japan

Hiroaki Fukada
Department of Mechano-Informatics, Graduate School of Information Science and Technology, The University of Tokyo, Japan

Taketoshi Mori
Department of Life Support Technology (Molten), Graduate School of Medicine, The University of Tokyo, Japan

Hiromi Sanada
Department of Gerontological Nursing/Wound Care Management, Graduate School of Medicine, The University of Tokyo; Tomomasa Sato, Department of Mechano-Informatics, Graduate School of Information Science and Technology, The University of Tokyo, Japan

Tomomasa Sato
Department of Mechano-Informatics, Graduate School of Information Science and technology, The University of Tokyo, Japan

Printed in the USA
CPSIA information can be obtained
at www.ICGtesting.com
JSHW011432221024
72173JS00004B/769